Selling Out

Consuming Ourselves to Death

DAVID MODEL &
LESLEY MODEL

authorHOUSE®

AuthorHouse™
1663 Liberty Drive
Bloomington, IN 47403
www.authorhouse.com
Phone: 1-800-839-8640

First published by AuthorHouse 8/10/2010

ISBN: 978-1-4520-4318-0 (e)
ISBN: 978-1-4520-4316-6 (sc)
ISBN: 978-1-4520-4317-3 (hc)

Library of Congress Control Number: 2010910199

Printed in the United States of America
Bloomington, Indiana

This book is printed on acid-free paper.

<u>DEDICATION</u>

Any Man's death diminishes me,
Because I am involved in mankind.
And therefore never send to know for
Whom the bell tolls; it tolls for thee.

John Donne

Table of Contents

Introduction

*E*aster Island has provoked historical and archeological controversy over the true causes of the near extinction of its peoples and destruction of its ecosystem. In one of the earlier studies Jared Diamond, American biologist and physiologist, explains the demise of Easter Island and its inhabitants as an act of self-destruction based on greed and ignorance. Despite the recent criticisms of his theories as inaccurate, they still can serve as a parable to unveil the shroud concealing the cataclysmic fate toward which humankind is unwittingly marching as lemmings racing towards the edge of a cliff. Jared's version of the self-destruction of Easter Island will shed light on the consequences of the collective self-destructive behavior of the human species threatening life as we know it.

Easter Island is located at 27 degrees south of the equator or approximately 2000 miles off the coast of Chile. According to Jared Diamond's research, it was discovered by Dutch explorer Jacob Roggeveen on Easter in 1722. He observed a wasteland without a single tree or bush over ten feet high where the only wildlife consisted of small insects. Its outstanding feature was over 200 huge stone statues called moai some of which stood more than 65 feet tall and weighed up to 270 tons. Jared claims that the statues were carved in a single quarry and transported up to six miles to their final destination.

In Jared's account of the history of Easter Island, human activity can be recorded as far back as A.D. 400 to 700 based on radiocarbon dating and he estimated that the number of inhabitants was approximately 7000 based on the density of archeological sites.

The island was a subtropical, pristine forest which towered over a layer of shrubs, herbs, ferns and grasses. Palm trees grew up to 82 feet tall and 6 feet in diameter. They were ideal for transporting and elevating the statues, constructing large canoes which were used for fishing and for yielding an abundant supply of edible nuts as well as sap to make honey and sugar. The diet of the people living on the Island consisted of flightless geese, pigs, dogs, porpoises, fish and a variety of seabirds. Rope was fabricated from the hauhau tree.

Jared believed that the inhabitants were colonists from Eastern Polynesia who set foot on a paradise with fertile soil, abundant food, a large supply of building materials, all of which held out the promise of a comfortable future.

After a few centuries, the inhabitants began carving, transporting and erecting stone statues for ancestral worship. Over time, the statues became larger and larger and more ornate. Transporting the statues demanded large quantities of wood and rope and as the number and size of the statues increased, the inhabitants had to cut down an ever-increasing number of trees. Over-cutting of palm trees and ubiquitous rats resulted in their ultimate extinction in the fifteenth century.

Wood was used to build canoes, fires, houses, and rollers for moving statues. Over-cutting of trees on Easter Island produced a number of crippling consequences which accelerated the destruction of the entire ecosystem. Trees soaked up the rainwater and their gradual disappearance caused soil erosion resulting in a loss of land for growing crops. Without wood to build canoes, the people of Easter Island were no longer able to catch fish and porpoises. Overexploitation of animals and birds at a rate exceeding their ability to reproduce doomed the inhabitants to the depletion of one of their major sources of food.

With the depletion of all the Island's sources of food and building materials, social order began to break down as their centralized government was replaced by a warrior class. Islanders turned on each other in a competition for increasingly scarce resources. By over-consuming their once abundant resources and destroying their environment, the population of Easter Island transformed a once abundant paradise into a wasteland which could barely sustain the few remaining survivors who were discovered by Jacob Roggeveen.

There is no mystery as to why the inhabitants did not realize the devastation they were wreaking on themselves. The pace of the destruction was so slow as to be imperceptible to people who were going about their daily business as usual. The overall pattern of interaction of the inhabitants with their environment was outside the scope of their daily preoccupations and as each new problem arose, they focused their attention on its solution. There was no external, objective voice to inform them of the destructive nature of their behavior and as their fate became clearer and clearer, social order had broken down and there was no mechanism to reverse the apocalyptic conditions under which they were now forced to live.

As the inhabitants on Easter Island faced a number of increasingly severe crises, so are the inhabitants of planet earth. The overall pattern is only visible to a few and the external, objective voices warning us about the consequences of our behavior have been marginalized to a significant extent, particularly in North America.

The evidence of this blindness to our fate is evident in the continued pursuit of fossil fuels in North America. For example, the tar sands in Canada and the proposed new offshore drilling for oil and expanded use of coal in the United States clearly illustrate that our leaders are unwilling to adopt sustainable technologies and that the population is not sufficiently alarmed to protest vigorously against such dangerous policies.

Global warming is another example of a crisis which most people now recognize but seem to accept placidly in the face of a complete lack of action on the part of our governments. Of course, there are examples of improvements, but overall, the prognosis is negative.

There are many more environmental problems that are barely on the radar screen but have already reached hurricane force. The rapidly growing depletion of fresh water threatens many parts of the globe with a shortage of drinking water. Modern farming techniques demand large quantities of water for irrigation. Coca Cola, Pepsi, Nestle and many other multinational corporations are draining aquifers to sell bottled water and toxins are rendering many fresh water sources too dangerous to drink. Alarmingly, global warming is melting glaciers that feed hundreds of millions of people.

The oceans are suffering a multiplicity of perils. These include destruction of coral reefs which are the bottom of the food chain, destruction of phytoplankton which is a major carbon sink, depletion of a majority of species which inhabit the oceans and pollution from plastic bottles and numerous toxins.

Industries have disseminated tens of thousands of dangerous chemicals into the atmosphere, water and ground. Surprisingly, animals as isolated as polar bears in the Arctic have ingested toxins which are then stored in their fat cells. Shockingly, all human fat cells are now a chemical warehouse, a latent threat which could cause cancer, birth defects and other diseases at any time. We are literally ticking time bombs.

Deforestation is continuing at an ever greater pace depriving the planet of a carbon sink, an essential mechanism in the hydrologic cycle and a potential source of drugs for curing disease. At least one species of plant or animal becomes extinct every day, mostly in the rainforests, depriving the planet of its essential biodiversity. Every species plays a role in nature and human ingenuity with its seductive technological prowess can't fix nature once it's broken.

Further, a growing scarcity of resources has impelled a number of wealthy nations to become neocolonial or imperialistic powers in pursuit of either ownership of or control over vital resources. The United States has become the major imperial power seeking control over oil resources to maintain its global dominance both politically and economically. Iraq, Afghanistan, Nigeria, Ecuador, Sudan, Somalia have at one time been victimized by the United States because they were unfortunate enough to possess an abundance of oil.

In addition, developing countries with essential resources, cheap labour, or large markets have become targets of exploitation by multinational corporations with the assistance of local dictators such as Obasanjo in Nigeria. Forcing these countries to borrow money from the World Bank to develop infrastructure for mining or drilling for the benefit of multinational corporations has enabled these corporations to virtually steal their resources. When these developing countries face hardship in making their payments to the World Bank, they are directed to the International Monetary Fund where further loans are authorized but only under certain conditions. These conditions force the borrowing country to dedicate its entire economy to loan repayment at the grievous expense of its inhabitants.

To produce energy, countries needed to dam rivers which causes flooding and displacement of local populations. Canada is an example of a country that dammed all the rivers flowing into James Bay forcing the Cree to live on reservations with no ability to live their traditional lifestyle. Another method of securing fossil fuels was to engage in massive exploitation of poor countries such as Nigeria and Ecuador. Foreign oil companies stole the oil which belonged to the people in these countries and destroyed their farmland and fisheries in the process. Nuclear fuel, touted as safe by governments and industry, is replete with problems, not the least of which is the disposal of radioactive waste.

Furthermore, the depletion of resources leads to conflict and wars of aggression as countries compete for ever-diminishing reserves of oil, minerals and metals not to mention precious objects such as diamonds. The Democratic Republic of the Congo (DRC) is a salient example of a country torn apart because of its abundance of rubber, gold, diamonds, copper, uranium, cobalt and coltan which is an essential ingredient in cell phones. At one time, eight different countries were at war with each other inside the DRC while the United States funded two surrogate states, Uganda and Rwanda, to invade the Congo to purloin its resources.

Protecting and controlling the supply of vital resources has been the cause of many conflicts and wars of aggression. Lurking in the background is the growing sophistication and number of nuclear weapons in the United States and in other countries while many non-nuclear states pursuit their own nuclear ambitions. Proliferation of nuclear weapons has moved the Bulletin of Nuclear Scientist's doomsday clock to five minutes before midnight, indicating a high probability of a nuclear disaster. In addition to a nuclear disaster is the looming danger of biowarfare as major powers accelerate their research and development of biological weapons of mass destruction.

While manufacturing the products we demand, industries frequently use toxic chemicals which are often dumped into the environment. More than 100,000 synthetic chemicals have been invented since World War II and many have never been properly tested for safety to humans. For example, chemicals such as dioxin, a by-product of waste incineration, PCB's, used as a coolant in transformers, BFR's used as a fire retardant in many products, benzene, used to clean computer chips infiltrate water, food and the air we breathe, expose us to a number of potential health disorders.

Paradoxically, there is one common cause to all these problems, the perceived imperative for growing consumption. Buy, buy, buy has been the "buy"-word in North American culture. People are strongly encouraged to buy products they really don't need to fuel economic growth. Economic growth is considered to be essential to produce jobs and raise our standard of living.

Production of theses goods deplete resources, devour huge quantities of energy, permeate the environment with toxic chemicals and create a massive amount of waste. These crises are not factored in to the measurement of growth. Therefore, continually growing consumption is based on unsustainable economic growth.

There really has not been a distinction between sustainable and unsustainable growth and measurements of growth reflect this weakness. Growth is promoted without recognition of the problems that accompany it. These crises are treated as a separate problem.

Production leads to growth but production today lacks consideration for the damage to the environment as well as the human rights and suffering of people both inside and outside North America. In other words, the more we grow, the more damage we cause to the environment and the more suffering we inflict on others who are not members of the circle of people who benefit from growing consumption.

Therefore, the economic growth that people in industrialized counties are enjoying today is not sustainable. The inevitable result of unsustainable growth is to exacerbate all the crises described above.

In addition, despite the fact that unsustainable growth has generated enormous wealth in the United States since World War II, it has not been distributed equally or fairly. In fact, there is no correlation between growth and quality of life factors such as healthcare, quality education, living wages and a safe environment. Although the United has the highest GDP in the world, it ranks last among OECD countries in most quality of life categories such as healthcare, infant mortality, child poverty, quality of education and unemployment. Ongoing increases in growth since World War II have not fulfilled their touted objectives of social and economic progress.

To shift the paradigm from a culture of thrift, frugality and moderation after World War II, business needed a plan to persuade people that shopping was the path to salvation. Part of the plan for continually expanding consumption was to create an axiom that the economic health of the nation depended directly on economic growth. It was essential for the creation of new jobs and for raising the standard of living. It alone could guarantee prosperity for the nation.

Economic theories always requires the assumption that "everything else being equal". It is not really a science because the human element is critical to predicting outcomes and human nature can't be reduced to an equation. Although in abstract economic terms, growth makes sense since more production will presumably create more job and increase wealth.

Issues such as the power of the owners of production to thwart efforts of workers to improve their wages, benefits and working conditions and to exploit cheaper labor in developing countries has been absent from social and economic discourse. So have discussions about the environment and depletion of resources except on very limited terms. Healthcare and other social programs to assist those experiencing hardship were also absent. So far, the election of Obama has changed very little. His healthcare plan is well short of a single-payer system which provides universal, accessible and affordable medical care. The financial reform package essentially neglects to restore the Glass-Steagall Act or control the wild casino-type investments which caused the disaster in the first place.

In fact, the most important economic measurement, the Gross Domestic Product (GDP), is the putative all-encompassing indicator of economic performance, masquerading as the true measure of the well-being of people in an economic system. GDP is a very useful device for convincing the public that if only we maintain a healthy growth rate, everyone will benefit. Yet, nothing could be further from the truth. For example, the Exxon Valdez, a 10.8 million gallon oil spill in Prince William Sound off the coast of Alaska, resulted in a total cost for cleanup, salvage, legal proceedings etc. of over $4 billion. This very destructive accident added over $4 billion to the GDP but this addition to the GDP was never understood in negative terms. It was simply, economic growth.

The alternative to the GDP, called the Genuine Progress Indicator (GPI), has not been adopted because while the GDP has been rising over the last 30 years, the GPI has been slipping for the simple reason that it incorporates human-related factors which really contribute to human well-being. No political or business leader wants to explain government policies as the product of corporate pressure and that corporations account for much of the downward movement of the GPI.

Reinforcing the myth about growth and social and economic health, presidents and economic advisors have been emphasizing the absolute necessity of ensuring economic growth since World War II. Every president since Eisenhower has either reassured Congress about economic growth or promised to promote it. Advisors, such as the Chairperson of the Economic Council of Advisors, have been heralding the benefits of growth thus further establishing the imperative of growth as a presupposition beyond questioning.

Inexorably growing unsustainable consumption is not an ideological, moral or judgmental issue but rather a pragmatic one. Pragmatism refers to the imperative to modify our lifestyles, expectations and priorities in order to rescue the ecosystem and human species from a catastrophe that could, in the worst case scenario, make life almost unbearable, if even possible.

The imperative for reducing unsustainable consumption and addressing all the above crises is extremely urgent given that only a small fraction of the world's population is currently responsible for these crises while the remainder of the population aspire to our lifestyle. Approximately one billion people live in advanced industrialized countries while over two billion people in China and India are desperately trying to surpass the United States in wealth. With three billion people consuming at the same rate per capita as the United States, there would be no hope of saving ourselves and the planet.

Compulsive buying did not infiltrate our culture by accident but was the result of a deliberate strategy after World War II to sustain ever-increasing corporate profits. In order to continue to fill the insatiable appetite of corporations for profits, the public would have to be persuaded to spend more of their income on goods and services. Artificially stimulating demand became an imperative to persuade the public to spend well beyond filling their basic needs. Three of the strategies developed to boost demand included planned, perceived and technological obsolescence.

Planned obsolescence involved manufacturing products that would be deliberately designed to have limited durability forcing consumers to replace them more frequently. If cars were designed to last a long time, then automobile manufacturers would quickly run out of customers. Inferior durability in cars was good for business as long as the consumer didn't lose faith in the manufacturer.

Business needed consumers to replace products before they were no longer unusable and devised a gimmick to convince the public that they needed to purchase a newer version of a product despite the fact that the older version was still in good condition. To achieve this objective, they would need the buyer to perceive that the older version was obsolete. By modifying style, fashions, accessories and other accouterments, consumers could be convinced that their older version was unsatisfactory or unacceptable and they would feel compelled to buy the latest model. This gimmick is called perceived obsolescence.

The advent of the computer and in particular, miniaturized circuits opened the floodgates to a multiplicity of products that constantly needed upgrading due to the accelerating advancements in technology. Cell phones, blackberries and computers head the list of products which are being rapidly rendered obsolete in the mindset of the consumer who is lured into replacing their older electronic equipment in order to enjoy the greater power and diversity of newer products.

Underlying the massive replacement of older products with new ones is the quickening pace of depletion of resources, pollution which results from the manufacturing and transportation involved in bringing the consumer goods to market, and the ultimate and inevitable disposal of these toxin-laden products.

Persuading consumers to open their wallets to replace ostensibly obsolete electronic products is referred to as technological obsolescence. The newer products are so much more powerful with such magnificent features.

The plan to convince consumers to purchase products they don't need has been very successful in terms of generating unsustainable growth in the economy. Concomitant with this growth has been the accelerating depletion of resources, the poisoning of the ecosystem and a growing inequality in the distribution of wealth.

While there is some strong evidence that we have begun to re-evaluate our relationship to consumption in recent years, this is mainly discernible vis-à-vis the marketplace with the mainstreaming of social, ecological and human rights issues. A collective understanding regarding the need for change in our consumptive patterns is, nevertheless, clear with the growing popularity of 'ethical' brands (Green products, Fair Trade etc...), as is business' desire to appease the social conscience of consumers with approaches to production. What remains to be seen is whether the rise of business' identification with Corporate Social Responsibility and the public's engagement with more 'ethical' forms of consumption can constitute the beginning of a substantive paradigm shift away from an unrestrained production/consumption model that privileges economic growth over the well being of all life forms on earth.

CHAPTER ONE

Grow Up:
How Economic Growth Became a Religion

Before the economic crisis greatly reduced people's disposable income, consumers were engaged in an orgy of frenetic buying in order to be seen in the right clothes or to disport themselves with the latest electronic gadgets. There is nothing intrinsically or morally wrong with indulging oneself in a life-long buying spree since the problem lies not with the individual consumer but with the invisible overall impact of consumption on ourselves and the planet.

Manufacturing, transporting, providing energy for and disposing of consumer goods depletes resources, pollutes the environment, generates waste and requires energy. The cumulative impact of world-wide consumption involving all these phases in the production cycle has had a devastating and deleterious effect on the planet and on all living species. For example, climate change, toxic chemicals in the air, food and water, destruction of the ocean ecosystem, destruction of the rain forests, a growing shortage of drinking water, loss of biodiversity, conflicts over a declining inventory of resources, and human suffering are only some of the results of manufacturing the products which people believe they need. These effects will be discussed further in another chapter.

The disastrous consequences of consumption beg the question as to why people feel the need to consume so much. Is it part of human nature, is it taught or a combination of both? Western culture profoundly etches into our psyches the need to define ourselves in terms of material possessions and to blur the distinction between needs and wants. Furthermore, we have been conditioned to ignore the overall consequences of consumption.

North Americans are so seduced by new technologies and the huge benefits they offer, the temptation to buy the dazzlingly sophisticated new products becomes irresistible.

For example, when cell phones first appeared on the market, people bought them for use in the case of an emergency but also as an extension of their land line. They soon became attached to people's ears. Then advancement in cell phone technology converted cell phones into computers, cameras, and television sets. Now, it carried all your electronic gadgets in one small electronic device that could fit into your pocket.

Next, blackberries appeared on the market tempting consumers with all the features of cell phones but also a GPS system, internet capabilities and more, serving as the new all-purpose electronic device. Blackberries expanded people's communication capabilities to such a high level that they could sit in the same room with someone and send them a text message. All these electronic methods of electronic communication are destroying real human to human exchanges. In the not-too-distant future, I can envision a new addiction group, Blackberries Anonymous.

As well, the antediluvian concept of maps for navigating to your destination was becoming unnecessary. Rather than waste time and brainpower reading a map, people can now watch the GPS system on their dashboard or blackberry and drive to any location in North America.

Another example, television, underwent many technological advances and on each occasion motivated consumers to avail themselves of the latest technological wizardry. When Beta video players became available, movie fanatics could purchase one and watch movies at home. Then VHS replaced Beta as the technology of choice forcing movie buffs to replace their Beta players and videos. But it didn't end there. DVD's replaced VHS as the latest technology for showing movies. Now it's HD DVDs. Every time a new technology replaced the old one, people who collected movies had to dispose of the old device and buy the new one and then also had to replace all their movies.

It is true that the more sophisticated and advanced devices are a greater convenience and offer more benefits but the consumer is not conscious of the overall impact of these products on the environment. The problem with consuming at current levels is its overall impact, not a reflection of the consumer.

Depletion of resources, pollution, and waste were not factors that had a noticeable impact on people's purchasing habits. For instance, when you buy a cell phone, you are not remotely aware of the six year old who mined the coltan in the Congo, the theft of that country's resources, or the death of millions of Congolese as countries scramble to purloin its resources. Nor are you aware of the women assembling the cell phone in a sweatshop who are exposed to dangerous chemicals. There are also energy costs in transporting the cell phone to the distributing centre in North America. Ultimately, the cell phone will become garbage at which point the toxic chemicals locked in the electronic parts will be released into the atmosphere or groundwater.

By continuing to buy new products, consumers guarantee increasing consumption thus satisfying the underlying economic imperative of growth in production on which prosperity is speciously based in most western economic systems. Growth has not really brought prosperity to everyone, the proof of which is the growing inequality of wealth in North America and the small percentage of people and corporations who benefit from expanding wealth.

Maintaining growth was industries' plan to ensure a steadily growing level of consumption so that their singular objective, the maximization of profits, would be realized on an ongoing basis. The fear that consumers would only buy what they needed would be alleviated through a plan to cultivate artificial demand for products through subterfuge.

Our impulse to devote so much of our disposable income or borrowed money obtained through credit cards or lines of credits is not a natural phenomenon but the result of a deliberate strategy of the government and business to ensure a constant growth in profits.

The crusade to guarantee a never-ending growing economy began after World War II when one of the major architects of American foreign policy, George Kennan, stated on February 24, 1948, that:

> We have about 50% of the world's wealth but only 6.3% of its population. We can not deceive ourselves that we can afford the luxury of altruism and world benefaction. We should cease to talk about such vague and unreal objectives as human rights, the raising of living standards and democratization. The day is not far off when we are going to have to deal in straight power concepts.
> (*Review of Current Trends in U.S. Foreign Policy*, George Kennan)

Kennan was saying that in order to maintain and increase the wealth of the United States, the U.S. government must be prepared to use force if necessary to protect America's access to natural resources, cheap labour, foreign markets and opportunities for investment. According to Kennan, wealth, or in other words, production, is America's number one priority not human rights, democracy nor the standard of living in other nations.

To persuade people to continually consume indefinitely, a number of economic and political leaders defined the ongoing acquisition of a growing number of products as the path to a happy and successful life. The philosophy that consumption was the path to fulfillment and a rewarding life was articulated by economist Victor Lebow in his paper *Price Competition in 1955* in the Journal of Retailing when he argued that:

> Our enormously productive economy... demands that we make consumption our way of life, that we convert the buying and use of goods into rituals, that we seek our spiritual satisfaction, our ego satisfaction, in consumption...We need things consumed, burned up, worn out, replaced, and discarded at an ever increasing rate.

Arthur F. Burns, Chairmen of the Council of Economic advisors under President Eisenhower, exceeded Lebow's orgasmic faith in the wonders of growth when he evangelized that: "The American economy's ultimate purpose is to produce more consumer goods." If Burns could observe the behavior of consumers today, he would feel a profound sense of satisfaction in the fact that consumers have reached Nirvana.

One year before Lebow proclaimed the sanctity of consumption, Brooks Stevens, an industrial designer, elucidated the method by which growing consumption could be sustained indefinitely. At a meeting of advertising agency executives in Minneapolis, he coined the phrase "planned obsolescence" although he did not invent the concept. He defined "planned obsolescence" as "instilling in the buyer the desire to own something a little newer, a little better, a little sooner than is necessary". (*Industrial Strength Design: How Brook Stevens Shaped your World*, Glen Adamson) Although Stevens' statement was controversial at the time, the various forms of obsolescence have become the backbone of indefinite economic growth.

Imbuing the imperative of growth with scholarly legitimacy, John Maynard Keynes, a British economist whose ideas were a major influence on modern economic and political theory until the 1970s, formulated a method to maintain growth despite downturns in the economy or even recessions.

One of his main contributions to economics was the principle that government has the power to increase production, incomes and jobs through government deficit spending. According to Robert B. Reich, Secretary of Labour under former President Bill Clinton:

> Keynes' basic idea was simple. In order to keep people fully employed, governments have to run deficits when the economy is slowing. That's because the private sector won't invest enough. As their markets become saturated, business reduce their investments, setting in motion a dangerous cycle: less investment, fewer jobs, less consumption.
>
> (*John Maynard Keynes*, Robert B. Reich)

Keynes believed that in times of slow business investment, it was the government's responsibility to engage in expansionary deficit spending to increase consumption by borrowing money and spending it. Although Keynes was not advocating unfettered growth, he reinforced the notion that consumption was the key to solving a number of economic problems.

Despite the fact that Keynes theories were discredited in the 1970s by the new rising star in economics, Milton Friedman, growth remained a key objective in government policy-making.

Milton Friedman, Nobel Prize winner and leader of the Chicago School of Economics, refuted Keynes theories to a large extent based on Keynes belief in the positive role of government in economic decision-making. Friedman was a strong advocate of the primary role of the market in making economic decisions. Although he disagreed with most of Keynes theories, he also believed in the importance of growth. According to Friedman:

> What kind of society isn't structured on greed? The problem of social organization is how to set up an arrangement under which greed will do the least harm; capitalism is that kind of a system.
>
> (*Capitalism and Freedom*, Milton Friedman)

From an economist standpoint fifty years ago, the main parameters in economic theories, apart from growth, were unemployment, inflation, interest rates, the money supply, trade, deficit, debt, trade, and the current accounts balance. Monetary and fiscal policy were designed to encourage the maximum growth in the economy to promote employment and a higher standard of living while minimizing the risk of high inflation or deflation or damage to the value of the dollar.. In the 1950's there was virtually no awareness about the negative impact of government economic policies on people and the environment and the ravages of growth.

The first hint that a major resource was approaching a limit was when M. King Hubbert predicted in 1956 that United States oil production would peak between 1965 and 1970. Peak oil means that the maximum rate of production has been reached pursuant to which the rate of production enters a terminal decline. Peak oil in the United States occurred around 1970. Many oil-producing nations have already passed their peak such as Venezuela, Libya, Iran, Nigeria, Indonesia, Oman, and Mexico. Although the science of estimating when oil production reaches its peak is controversial, Hubbert's prediction should have set off some alarm bells.

Another warning was issued when Rachael Carson wrote *The Silent Spring* in 1962 in order to call attention to the dangers of pesticides on human health and in particular, DDT. The Chemical industry lambasted her in order to destroy her credibility but environmental awareness was hatched.

Then in 1972, the crisis of growth was recognized by a group of scientists who formed an organization called "The Club of Rome". Phase One of the project was conducted at MIT which:

> Examined the five basic factors that determine, and therefore, ultimately limit growth on this planet – population, agriculture, natural resources, industrial production, and pollution...Our conclusions from these extrapolations is...that the short doubling times [of exponential growth]

of many of man's activities...will bring us closer to the limits of growth of these activities surprisingly soon...The crux of the matter is not only whether the human species will survive, but even more whether it can survive without falling into a state of worthless existence.
(*The Limits to Growth*, Donella H. Meadows, Dennis Meadows, Jorgen Randers, William W. Behrens III, p.11)

Since 1972, there has been a steady stream of literature warning governments that continually ignoring the dangers of unsustainable growth is risking human life and the ecosystem. One of the crucial warnings was that the atmosphere, water supply, and land are not infinite resources and that continually treating them as such is short-sighted and foolish.

Today, there are a myriad of environmental organizations demanding that governments rethink their priorities when formulating social and economic policy. Tragically, growth is still virtually unsustainable and persists as the number one priority in economic policy, the proof of which is the ongoing pursuit of fossil fuels and the absurdly inadequate funding for adopting alternatives.

One of the important factors contributing to the notion that growth was the key to a strong, healthy economy was the litany of presidential economic advisors who strongly advocated for the miraculous powers of growth.

The Council of Economic Advisors is an agency of the executive branch of government consisting of three respected economists who advise the president on economic policy. It provides much of the economic policy for the White House. Moreover, the Council of Economic Advisors submits a report on the economy to the president every year. Without exception, these advisors were fully committed to expansionary economic policies to lower unemployment and raise the standard of living. The problem is that they were neither aware of nor cared about unsustainable growth. In fairness to these advisors, in the first decades after World War II, there was virtually no awareness of the concept of sustainability.

For example, in the 1953 Report, Arthur F. Burns was its Chairman at the time, reveals the firm commitment to maintaining economic growth when Burns stated that:

> I believe that our resources are sufficiently plentiful to reach a 1953 national product as high or even higher than that set out in this review. The idea, often expressed by this Council, and in fact by other persons through the ages, that a free and great people should be able to use all of what they have the ability to produce seems to me so commonplace... This issue- whether we shall shrink from or measure up to the challenge of potential abundance-is perhaps the supreme issue of the Twentieth Century.
> (*Annual Economic Review*, The Council of Economic Advisors)

Excerpts from the reports of later Chairman reflect this commitment to unsustainable growth as evidenced by the following examples:

- 1961 - Raymond J Saulnier under Eisenhower: "Throughout 1958, Government expenditures for goods and services continued to rise fairly steadily...The expansion of the economy proceeded vigorously, supported by steadily mounting consumption."
- 1965 - Gardiner Ackley under President Johnson: "A wide consensus of responsible opinion now recognizes that Federal fiscal policy must be geared to keep the economy moving ahead. It [Annual Report] attempts to dissect the unprecedented long and healthy expansion of the past four years."
- 1973 - Herbert Stein under President Nixon: "There was a rapid acceleration in the rate of growth in real output...The rise in production and employment that became particularly evident by midyear created a growing mood of confidence...and helped to bolster private investment and personal consumption."
- 1977 - Charles Schulter under President Carter: "Strong and steady growth in the U.S. economy was also needed to help sustain the pace of economic expansion among the nations of the Western World."

- 1986 - Beryl W. Sprinkel under President Reagan: "The outlook is favorable for continuation of a healthy expansion. After slowing in the second half, economic activity is again accelerating."
- 1995 - Laura D'Andrea Tyson under President Clinton: "By most standard economic indicators, the performance of the U.S. economy in 1994 was, in a word, outstanding. The economy has not enjoyed such a healthy expansion of strong growth and modest inflation in more than a generation."
- 2007 - Edward P. Lazear under President George W. Bush: "The U.S. economy continues to exhibit robust growth...Much of this Report explores the role of productivity-related issues in the continuing expansion of the U.S. economy."
  (*Annual Economic Review*, The Council of Economic Advisors)

Note the use of the words "vigorously", "healthy", "rapid", "strong", "accelerating", and "robust" to describe, what was, in fact, unsustainable growth. Since there were no distinction drawn between sustainable and unsustainable growth, it is very clear that the only indication of a healthy economy was growth of any kind. The leading economic advisors to the president were not at all aware of nor concerned with sustainability but only in growth to promote employment and prosperity. It is possible to create jobs and improve the standard of living with sustainable growth by transforming the economy from a fossil fuel, toxic polluting and disposable system to one that is people and earth friendly.

Reinforcing the crucial nature of growth, without any acknowledgement that it has two paradigms, sustainable and unsustainable, is the economic analysis to be found in the presidents' annual State of the Union Address to Congress in which he reports on the condition of the nation and outlines his legislative agenda. As the key measurement of the performance of the economy, growth is given considerable attention in the Address. Every President since World War II has treated growth as the touchstone of the success of his economic policies. For example:

- Dwight Eisenhower (1958): "There are solid grounds for confidence that economic growth will be resumed without an extended interruption."
- John F. Kennedy (1963): "While programs for full utilization of existing resources are the indispensable first step in a positive policy for faster growth, it is not too late to move ahead on other programs to strengthen the underlying sources of the nation's capacity to grow."
- Lyndon Johnson (1965): "Our basic task is threefold: First to keep our economy growing."
- Richard Nixon (1970): "Our gross domestic product will increase $500 billion in the next years."
- Jimmy Carter (1978): "First the economy must keep on expanding to produce new jobs and better income, which our people need."
- Ronald Reagan (1984): "The key to a dynamic decade is vigorous economic growth, our first real goal."
- George H. W. Bush (1991): "We will get on our way to a new record of expansion and achieve the competitive strength that will carry us into the next American century."
- Bill Clinton (1997): "Over the last 4 years, we have brought new economic growth."
- George W. Bush (2002): "The way out of this recession, the way to create jobs, is to grow the economy."
- Barrak Obama (2009): "Now is the time to act boldly and wisely – not only to revive this economy, but to build a new foundation for lasting prosperity. Now is the time to jumpstart job creation, re-start lending, and invest in areas of energy, health care, and education that will grow our economy."
  (*Speeches by US Presidents*, State of the Union Address Library)

In Barrack Obama's address, there is a section on energy in which he promises to reduce greenhouse emissions and to introduce more alternative forms of energy. There seems to be a recognition that the energy supply for growth and transportation should rely more heavily on sustainable fuels. According to his report:

> Thanks to our recovery plan, we will double this nations supply of renewable energy in the next three years. But to truly transform our economy,..We need to ultimately make clean, renewable energy...So I ask this Congress to send me legislation that places a market-based cap on carbon pollution and drives production of more renewable energy in America.
>
> (*Speeches by US Presidents*, State of the Union Address Library)

Obama's cap-and-trade plan is not only misleading, it is seriously flawed. It is misleading because payment of a fee to the government by coal users will result in higher prices for coal products and will be equivalent to a regressive tax on energy consumption. *Business Week* claims that:

> Obama proposes that companies buy an allowance, or permit, for each ton of coal emitted, at an estimated cost of $13 per ton to $20 per ton. ... Energy companies and utilities would likely pass along the added cost to consumers. It's estimated the price of gasoline would go up 12 cents a gallon and the average electricity bill by about 7%.
>
> (John Carey, Business Week, March 5, 2009)

The *Wall Street Journal* reports that.

> The Congressional Budget Office...estimates that the price hikes from a 15% cut in emissions would cost the average household in the bottom income quintile about 3.3% of its after-tax income every year.
>
> (Who Pays for the Cap and Trade?, The Wall Street Journal, March 9, 2009)

As well, President Obama's plan does not call for a phase-out of coal which releases more than twice as much carbon dioxide as oil into the atmosphere. Reducing emissions from coal by 15% over the next three years is a molecule in the bucket. Eighty-five percent of America's energy demand is met by fossil fuel and coal constitutes 45% of fossil fuel use.

Cap and trade will result in a 6.75% drop in fossil fuel usage over the three year period.

Nevertheless, climate scientists have concluded that it is imperative that a twenty-year phase-out in coal emissions within the next twenty years is essential to a real solution to greenhouse emissions because it constitutes 80% of the solution to climate change. Assuming a plan in which every three years the President commits to reduce emissions by a further 6.75% every three years, it would take twenty-one years simply to reach a 40 % reduction. His strategy falls considerably short of the timetable climate scientists have established for warding off the ravages of climate change.

Furthermore, Obama's plan to double the supply of renewable energy in the next three years is very misleading since renewables currently only account for 3% of U.S. production of energy. The climate change crisis requires much stronger action.

His plan to ask Congress for $15 billion for advanced alternative technologies reflects backward thinking in terms of safeguarding the planet from the impact of fossil fuels. In his address he states that:

> And to support that innovation, we will invest $15 billion a year to develop technologies like wind power and solar power; advanced biofuels, clean coal, and more fuel-efficient cars and trucks built right here in America.

(*Speeches by US Presidents*, State of the Union Address)

The promotion of biofuels is a step in the wrong direction for a number of reasons. Crops from which biofuels are made require petroleum products, large amounts of water and pesticides, all of which cause environmental harm greatly outweighing the benefits of producing the fuel. In addition, by removing so many crops from the international trade in food, the prices of corn, sugar cane and soy oil will increase on international markets resulting in more people who will drop into the sinkhole of extreme poverty.

Clean coal is another fantasy inhabiting the minds of Obama's policy-makers. It creates the illusion that the president is seriously addressing the dangers of fossil fuels while failing to neglect the virtually insuperable problems surrounding this technology. Coal will still have to be mined and currently one of the major sources of new coal is the Appalachian Mountains where the tops are blown off to gain access to coal. Apart from disfiguring the landscape, the process produces toxic wastes which are presently dumped into valleys allowing the chemicals to pollute the water system.

Liquefying coal to produce fuel releases huge quantities of carbon dioxide which have to be captured and stored and, at this point in time, no satisfactory technology has been developed. The costs of producing fuel from coal are an important issue and may render the entire process unfeasible. According to the Center for Sustainable Economy:

> The two CTL (coal to liquid) economic analyses completed to date demonstrate that it is highly unlikely that CTL development is viable from the public perspective because net present values are likely to be negative and benefit-cost ratios well below one, implying that CTL development will probably cause more economic harm than good.
> (*Economic Feasibility of coal to Liquids Development in Alaska's Interior*, John Talberth)

Hence, despite the fact that President Obama laid out a plan for renewable energy, he fails to really address the problem of fossil fuels. That means that his use of the word growth still means unsustainable growth. The word sustainable never appears once in his address to Congress whereas current thinking about the economy needs a paradigm shift from unsustainable growth to sustainable growth.

A large part of the problem is that the connection between sustainable and growth does not exist in the minds of American leaders past and present. Growth is considered to be vital to the health of the economy and the well-being of Americans but there is no distinction between the two types of growth. In public discourse, the term growth must be replaced with sustainable growth.

One of the critical problems is that the measurement of growth, the Gross Domestic Product, fails to account for sustainability and therefore its value is flawed. If sustainability were built into the measure, the economic picture would be quite different.

CHAPTER TWO

Measuring Up:
Growth Measurements are Misleading

*I*n *2005, there were 6,420,000* automobile accidents in the United States resulting in the deaths of 42,636 people and the injury of 2.9 million. Surprisingly, the total financial cost of these accidents was $230 billion. (Car Accident Statistics, Car-Accidents)

Is there any conceivable positive interpretation of this data and in particular, an interpretation that would suggest that these accidents resulted in progress in any sense in the United States? Silly question? Shockingly, according to the most important economic indicator used by policy-makers in the United States and around the world, the gross domestic product (GDP), these car accidents contributed to economic progress in the United States. Progress as measured by economic growth did increase. This strange anomaly is only one of many flaws in the GDP that render it meaningless as a measure of progress of any kind other than economic transactions. In fact, many of the important economic indicators on which social and economic decisions are based such as the unemployment and poverty rate are also quite deceptive and misleading.

The main problem is that there has been no distinction between the two kinds of growth and furthermore, growth was considered healthy without recognizing the problems with unsustainable growth or methods of production.

Sustainable growth can be subdivided into three categories, economic, social and environmental. Sustainability is commonly associated with environmental sustainability referring to management of resources, minimizing pollution and reversing or halting all the catastrophic trends such as global warming and deforestation.

Social sustainability involves achieving social justice by reducing inequality and eliminating poverty. These social issues are fraught with dangers as well, if they are not addressed. Apart from the moral imperative to reduce inequality and eliminate poverty, stability and order are at risk if we ignore the disadvantaged.

To create a sustainable economy, government deficits and debts must be managed so that the economy is viable and can meet the needs of the people without burdening future generations with an enormous debt. It also means managing international trade so that the current account balance is not allowed to become another burdensome debt which will eventually come home to roost. As well, taxes and government expenditures must be sustained in some sort of equilibrium so that social programs and services are affordable and taxes are not too high.

Since there is still no distinction between sustainable and unsustainable growth, according to the GDP car accidents contribute to so-called progress along with forms of growth that are, in fact, unsustainable.

The GDP is defined as the total value of all goods and services that are produced and traded for money in a given time period in a particular nation. For example, it measures the total dollar value of all the automobiles, houses, toasters, furniture, banking services, medical services, and legal services etc. that are produced and then exchanged for money.

In 1934, Simon Kuznets, head of the U.S. national accounting system, introduced the concept of the gross domestic product as a measure of economic growth or activity but not as a measure of progress in any other sense. Over the years it morphed into a measure of overall progress despite the fact that it does not measure anything other than the production and exchange of goods and services. It is now used by economists, the business community, politicians and international institutions such as the World Bank, International Monetary Fund, United Nations and the Organization of Economic Cooperation and Development as the main indicator of economic health.

The dangers posed by the GDP proliferate when you consider all the other indicators that use the GDP as a basis for comparison for the purpose of international comparisons. Both total taxes and total debt are compared to the gross domestic product in order to compare one year to another or one nation to another. As well, in order for a comparison among nations to make sense, measurements must be calculated on a per capita basis.

When the U.S. government speaks of growth, it is referring to an increase in the GDP from one month to another or one year to another. Therefore, progress in any nation means that more goods and services were produced and sold in one period compared to the previous one. It is clear that the exclusion of sustainability in the calculation of the GDP is a serious flaw but not the only one. The flaws in the GDP are:

- Negative counting: Car crashes fall into this category in which negative occurrences generate economic activity but are negative events which by no means should be treated as progress. Included in this category are cigarette smoking, oil spills, divorces, defense spending, prisons, disasters and fires. Defense spending alone accounts for approximately 20% of the GDP.
- Excluded activities: Activities such as volunteer work, costs of crime, prostitution, depletion of natural resources, costs of damage to the environment are not counted in the GDP.
- Who benefits: Growth means an overall increase in wealth but the GDP does not reflect who benefits from the increase. There

has been a growing gap between the rich and the poor in the U.S. The New York Times reports that

Average incomes for those in the bottom 90 percent dipped slightly compared with the year before, dropping $172 or 0.6 percent. The gains went largely to the top 1 percent, whose incomes rose to an average of more than $1.1 million each, an increase more than $139,000, or about 14 percent. The new data shows that the top 300,000 Americans enjoyed almost as much income as the bottom 150 million Americans. Per person, the top group received 440 times as much as the average person in the bottom half earned, nearly doubling the gap since 1980.

(*Income Gap is widening*, Data Shows, David Cay Johnston)

- Depletion of Resources: Ironically, the faster resources are depleted, the higher the GDP rises. In fact, GDP encourages depletion of resources because the value of the trees that are cut down will be more valuable economically then the loss of bio-diversity, loss of habitat, flood control, and a carbon sink. These factors are not included in GDP calculations. The trees will be sold as lumber and as finished products or used in construction, all of which will contribute to the GDP whereas the other factors will not.

- Quality of life: Such important elements as quality of education and healthcare, freedom, privacy, security, and leisure time are not part of the GDP formula.

- CPI Deflation: In order to compare the GDP over time, it is critical to remove the inflation factor or, in other words, the increase in GDP resulting from rising prices as measured by the Consumer Price Index (CPI). Inflation increases the value of the GDP without representing any increase in economic activity. In 1996, the formula for calculating the CPI was changed by altering the weighting system so that the CPI would be revised downward. If the CPI is smaller, the deflation factor applied to the GDP is smaller creating the illusion that the economy is growing faster than economic activity would warrant.

It is clear from the above flaws that a country can have a high GDP while, at the same time, there may be a lack of freedoms, basic services and opportunities to enjoy a fulfilling life. A prime example is the United States which has the highest GDP in the world in absolute terms but ranks tenth in per capita terms with the following countries ranking lower:

GDP per Capita 2009

COUNTRY	PER CAPITA GDP	RANK
U.S.	$46,400	8th
Netherlands	$39,200	16th
Canada	$38,400	19th
Sweden	$36,000	20th
Belgium	$36,600	22th
Denmark	$36,000	23th
UK	$35,200	24th
Germany	$34,100	26th
France	$32,800	28th

(*Country Comparison – GDP-per capita*, Central Intelligence Agency <CIA> Fact Book)

Based on the assumption that the GDP really measures progress and well-being, it would only be reasonable to assume that major factors contributing to progress would have some correlation with gross domestic product. Despite the fact that the U.S. ranks ahead of the above countries in GDP, it ranks either at the bottom or near the bottom in a number of categories relating to progress and quality of life.

Consider the following economic and social indicators:

Child Poverty in Rich Countries – 2005

Country	Child Poverty
Denmark	2%
Sweden	4.2%
France	7.5%

Belgium	7.7%
Netherlands	9.9%
Germany	10.2%
Canada	14.9%
UK	15.4%
US	21.9%

(*Country Report No. 6*, UNICEF)

Not only is the United States ranked last among these countries but amazingly, its child poverty rate is double the first six countries on this list.

Unemployment rate - 2008

Country	Unemployment
Denmark	2%
Netherlands	4.5%
UK	5.5%
Canada	6.1%
Sweden	6.4%
Belgium	6.5%
US	7.2%
France	7.4%
Germany	7.9%

(Unemployment Rate, CIA World Fact Book)

Infant Mortality Rates
Per 1000 live births
2009 (estimated)

Country	No. Deaths
France	3.33
Germany	3.99
Denmark	4.34
Belgium	4.44
Netherlands	4.73
UK	4.85

Canada   5.04
US   6.26
Sweden   N/A
(Infant Mortality Rates, CIA World Fact Book)

Health Care Ranking - 2000

Country	Ranking
France	1
Norway	11
Netherlands	17
UK	18
Belgium	21
Sweden	23
Germany	25
Canada	30
US	37

(Ranking of Worlds Healthcare systems, World Health Organization)

In one case, the United States was third last and in all the other cases, it was dead last despite the fact that it had the highest GDP of all the countries to which it was compared. Clearly a high GDP is no guarantee of progress or well-being. To use it as such is very misleading and misrepresents the health of the economy and the welfare of Americans.

Ironically, there are a number of very credible alternatives to the GDP which actually do reflect the progress of the nation in many respects, not just in economic activity. These measurements incorporate such factors as literacy, education, distribution of income, sustainable use of resources, sustainable use of ecosystems, cost of water and air pollution, leisure time and cost of underemployment. These indicators not only measure sustainable growth but a number of quality of life fundamentals.

For example, the United Nations developed the Human Development Index (HDI) which does incorporate the Gross Domestic Product in its formula but also:

> Looks beyond GDP to a broader definition of well-being, living a long and healthy life (measured by life expectancy), being educated (measured by adult literacy and enrolment at the primary, secondary and tertiary level)....What it does provide is a broadened prism for viewing human progress and the complex relationship between income and well-being. (*Human Development Reports: 2008 Statistical Update*, United Nations)

The HDI ranges from zero to one, with one representing the highest level of development. According to the HDI, a comparison of the nine countries above reveals that:

UN Human Development Index
2006

Country	Index
Belgium	.968
Canada	.967
Netherlands	.958
Sweden	.958
France	.955
Denmark	.952
US	.950
UK	.942
Germany	.940

(*Human Development Reports*, United Nations)

Again, the United States ranked seventh despite the fact that it had a higher GDP than all the other countries, further proving that GDP has no correlation with quality of life or progress.

Another alternative to the GDP is the Genuine Progress Indicator (GPI), created in 1995 by Redefining Progress, to measure how citizens are doing both socially and economically. It begins with the GDP but then adds and subtracts factors based on whether they contribute to or detract from the well-being of citizens. The following are some of the factors incorporated into the GPI:

- Personal Consumption: This amount is based on the GDP.
- Income Distribution Index: The extent of the inequality in income based on the Gini index is subtracted from the total. The Gini index ranges between zero and one and the higher the Gini index, the greater the income inequality.
- Value of Higher Education: This factor is partly based on the number of students in post-secondary education and the estimated dollar amount of social benefits of higher education.
- Cost of Crime: This amount is based on the psychological impact and lost opportunities resulting from crime. It is based on the Bureau of Justice Statistics National Crime Survey year to year estimates of the cost of crime to victims.
- Loss of Leisure Time: The loss is due to the fact that people are working harder to pay their expenses and to meet their debt obligations. The amount is based on an estimate of the average increased hours of work from year to year.
- Cost of underemployment: Underemployment refers to people who are unemployed, discouraged from seeking work after months or years of trying and those who work part-time but need and seek full-time employment. A study undertaken by Leete-Guy and Schor in 1992 estimates that the number of hours of underemployment in 1989 added up to $14.6 billion. In calculating the number of underemployment for any given year, the same method is applied and the total hours are multiplied by the average real wage per hour according to the Bureau of Labor Statistics.
- Loss of farmland: Data was combined from a number of sources including the American Farmland Trust, National Agricultural Statistics Service and National Agricultural Lands Study to estimate the average annual farmland lost to urbanization. The

cost of this land was based on its value according to the above studies.

- Depletion of Nonrenewable Energy Resources: Current accounting systems for the depletion of fossil fuels, treats the value of usage as income rather than capital. Since these resources are not renewable, their usage is a loss of capital. The GPI estimates renewable energy replacement costs as an approximation for the loss of nonrenewable energy sources. In other words, the cost is the amount of renewable energy investment that would be needed to replace the loss of the use of fossil fuel. The GPI relies on the cost of biomass fuel as the replacement source of energy because it represents 47% of the renewable energy market.

- Carbon Dioxide Emissions Damage: There is a wide consensus among scientists that there is a correlation between greenhouse emissions and climate change and also that there is a correlation between climate change and the increasing incidence of severe storms, floods and droughts. The damages from the above disasters escalate the cost of insurance payouts and replacing or repairing damaged homes and infrastructure. The GPI uses data on greenhouse emissions as reported by the Oak Ridge National Laboratory and the cost of the damage is based on a major study which combined the results of 103 separate studies.

These are only some of the factors incorporated into the GPI and transform the GDP into a meaningful measure of prosperity, well-being and progress. It is important to explain why the government has not adopted one of the many indexes that more accurately reflect progress in a nation. One possible reason might be lack of awareness of the new indices or it might be related to the problem of setting up a mechanism for calculating the new index. A more likely explanation might be understood in the context of the controversies surrounding the calculations needed to determine the value of the GPI or other comprehensive indices. An even more likely explanation is that politicians would be required to explain why the GPI is negative or grows more slowly than the GDP. Any honest explanation would need to address the depletion of resources, inequality of wealth and the failure to search more vigorously for alternative sources of energy and for a real solution to climate change. It is reasonable to assume that since politicians seem to lack the political will or motivation to solve these problems now, they would not want to switch to a measurement that might force them to do so in the future.

A comparison of the GDP and GPI since 1950, demonstrates the growing distance between the two which can be explained by a lack of attention to the problems identified in the factors incorporated into the GPI.

The following table compares the value of the GDP and GPI per capita between 1980 and 2004. The percentage change from one year to the next is also shown:

YEAR	GPI PER CAPITA	PERCENT CHANGE	GDP PER CAPITA	PERCENT CHANGE
1980	$14,730.24	N/A	$22,666.27	N/A
1981	14,682.31	- .33%	23,010.79	1.52%
1982	14,722.31	+ .27	22,349.56	-2.87
1983	15,231.57	+3.46	23,148.26	+3.57
1984	14,921.56	- 2.00	24,597.63	+6.26
1985	15,123.92	+1.35	25,386.01	+3.20
1986	15,122.17	0.00	26,027.73	+2.53
1987	14,960.25	- 1.07	26,668.01	+2.46
1988	14,913.77	- .31	27,518.87	+3.19
1989	14,967.38	+ .36	28,225.70	+2.57
1990	14,892.80	- .49	28,434.99	+ .74
1991	14,575.01	-2.14	28,010.64	-1.49
1992	14,342.57	-1.59	28,558.86	+1.96
1993	14,175.75	-1.16	28,943.54	+1.35
1994	14,051.40	- .88	29,743.47	+2.76
1995	14,409.10	+2.55	30,131.27	+1.30
1996	14,508.48	- .69	30,886.87	+2.51
1997	14,410.04	- .68	31,891.23	+3.25

1998	14,553.23	+1.05	32,837.40	+2.97
1999	15162.06	+4.18	33,907.88	+3.26
2000	15,145.93	- .11	34,764.23	+2.52
2001	14,417.04	- 4.81	34,665.17	- .28
2002	14,765.33	+2.41	34,866.86	+ .58
2003	14,807.16	- .28	35,460.01	+1.70
2004	15,035.65	+1.55	36,595.59	+3.20

(The Genuine Progress Indicator 2006; J. Talberth, C Cobb, N. Slattery)

Notice that the total growth in GPI per capita over the twenty five years is $305.41 while the total growth in GDP per capita over the same time period is $13,929.32. According to GDP, there has been 61.45% growth between 1980 and 2004 signifying, in the context of current interpretations of GDP, that the well-being of Americans and health of the economy have enjoyed substantial progress.

On the other hand, according to GPI, there has been 2.07% progress over the same period of time. The difference between the two indicators represents the exclusion of important quality of life, environmental and well-being factors which have virtually discredited economic growth as a reflection of progress.

The median growth rate for GDP between 1980 and 2004 is +2.51%. That means that half the growth rates for each year were above 2.51% and half were below. As well, the median growth rate for GPI over the same time period is +.28% which means half were above and half below +.28%. Therefore, when the GPI is used to measure progress during this 25 year period, half the growth rates were close to zero or negative suggesting very little progress or regression.

Another statistic, the increase in people's wages since 1970, confirms that most Americans have not enjoyed any improvement in their living standards over this period of time. The New York Times reports that:

In real terms, the wages of nonmanagement employees in the United States are now 10 percent below their level in the early 1970s, according to Labor Department Statistics.

(*After Years of Growth, What About Workers' Share?* Eduardo Porter)

Therefore, just in income alone, there has been a 10 percent decline in workers wages since the 1970s despite the 61.45% increase in GDP. Workers did not benefit from growth during that 25 year period. Also, there are still a number of other factors that could have been considered in analyzing the 61.45% increase, such as pollution and healthcare, which would have affected the quality of life of most Americans. Clearly, the 61.45% increase in GDP is very misleading and conceals a number of significant problems that need to be addressed.

Americans are now suffering from the outcome of policy choices based on GDP. Many families need two incomes to survive and have less leisure time. Effective healthcare is beyond the means of many Americans and approximately 20,000 people die each year due to unaffordable medical services. Water shortages are becoming a problem in a number of States while toxic chemicals are pervasive in the environment.

Clearly, unsustainable growth as measured by the GDP is not a useful method for measuring well-being of people within a nation. It is misleading and deceptive and creates the illusion of progress when, in fact, the reverse is often true. Both people and the environment suffer while the GDP suggests otherwise.

CHAPTER THREE

Work Up:
Growth Does Not Reflect the Plight of Workers

There are tens of millions of Americans who are working but unfortunately are not earning a sufficient income to pay all their bills. Yet, these Americans are invisible in the unemployment statistics even though they are struggling to feed their families. They belong to a category of workers more aptly defined as inadequately employed or underemployed. Dividing workers into adequately employed and inadequately employed yields a more accurate blueprint of the workforce in America.

As well, the rate of growth does not reflect the extent to which workers have benefited as a result of increasing wealth. As was pointed out in the previous chapter, growth fails to reflect the share going to workers. Hence, a high growth rate is misleading in the sense that it fails to reflect the viability of worker's incomes.

In addition, growth fails to reflect the ongoing loss of jobs due to outsourcing, technology, downsizing and replacement of full-time jobs with part-time jobs. This contraction of the job market is devastating to workers who are compelled to seek more than one job, work longer hours or work part-time to earn a sufficient income to pay all their living expenses. Most of the new jobs are low-paying, poor-benefit service sector jobs.

According to the U.S. Bureau of Labor Statistics, the official definition of unemployment means that:

> Persons are classified as unemployed if they do not have a job, have actively looked for work in the prior 4 weeks, and are currently available for work.
>
> (*How the Government measures Unemployment*, U.S. Bureau of Labor Statistics)

Unemployment rates for the years 2004 to 2008 are:

Year	Unemployment rate
2004	5.5%
2005	5.1%
2006	4.6%
2007	4.6%
2008	5.8%

(*Employment Status of the Civilian Noninstitutional Population*, Bureau of Labor Statistics)

There are four major groups that are not accorded the correct status in the unemployment rate: part-time workers who continue to seek full-time jobs; contingent workers who seek full-time jobs; the working poor who have jobs but are living below the poverty line and workers who have given up seeking employment altogether.

A worker is defined as part-time if they work 34 hours or less in a week. There are a number of reasons why a worker might work part-time such as:

- slack work or business conditions;
- unable to find full-time work;
- seasonal work;

- child-care problems;
- family or personal obligations;
- in school or training;
- weather-related problem.

Workers who are part-time but would prefer a full-time job (involuntary part-time workers) are counted as employed despite the fact that they may not be earning a sufficient income to support themselves and their family or further, need more than one job to do so. In 2008, 23.84% of all workers worked part-time and 17.6% of those workers were part-time workers who were working for an average of 22.9 hours per week but not by choice. (*Household Data: Annual Averages*, Bureau of Labor Statistics)

Therefore, in 2008, 4.2% (17.6% of 23.84%) of all workers were under-employed because they were involuntary part-time workers. The first step in calculating the percentage of workers who are underemployed is to add the unemployment rate (5.8%) to the percentage of involuntary part-time workers for a total of 10.0%.

Failing to secure full-time employment is not the only problem faced by these workers. They are also, on average, paid lower wages and enjoy fewer benefits. The mean hourly rate for full-time workers in the United States is $21.08 while the mean hourly rate for part-time workers is $11.02. (*Occupational Earnings Tables - 2007*, Bureau of Labor Statistics) Even averages mentioned above are misleading simply because they do not reflect the distribution of wages among part-time and full-time workers.

Another method for evaluating wages is to partition them into percentiles where a percentile is the value of a number (wage in this case) below which a certain percentage of all the observations fall. For example, the twentieth percentile means that twenty percent of all the numbers under consideration fall below the twentieth percentile. An examination of the breakdown of wages according to percentiles reveals that:

Part-time Civilian Workers: Hourly Wage Percentiles (2007)

Percentile	10%	25%	50%	75%	90%
Wage	$6.15	$7.25	$8.77	$12.00	$19.40

(*Part-time Civilian Workers*, Bureau of Labor Statistics)

Ten percent of all part-time workers earn less than $6.15 an hour and twenty-five percent of all part-time workers earn less than $7.25 an hour.

Full-time Civilian Workers: Hourly Wage Percentiles (2007)

Percentile	10%	25%	50%	75%	90%
Wages	$9.00	$11.71	$16.88	$25.83	$37.70

(*Full-time Civilian Workers*, Bureau of Labor Statistics)

There is a large gap between full-time workers and part-time workers in every percentile. In particular, in the 50th percentile, the maximum hourly wage for full-time workers is $16.88 twice that of part-time workers ($8.77). Therefore, there is strong evidence given a relatively even distribution of wages within each percentile, that ½ of part-time workers are earning ½ of the wages paid to full-time workers.

Despite the fact that full-time workers earn, on average, $21.05, half of these workers earn less than $16.88. On the other hand, while the average hourly wage for part-time workers is $11.34, half of these workers earn less than $8.77. There is a large difference between the average wage and the wage at the 50th percentile (medium wage) due to the fact that the average wage is highly susceptible to the huge wages at the very high end whereas the medium is not.

Lack of benefits is another predicament forcing low wage earners to pay for health care, save for their retirement and lose pay on sick or vacation days. A low wage earner is defined by scholars as:

> One in which a full-time year round-worker- someone who works 40 hours a week for 52 weeks or 2,080 hours in a year- earns less than the

poverty threshold for a family of two adults and two children ($20,444 in 2006, or $9.83 an hour).

(*Understanding Low Wage Work in the United States*, The Mobility Agenda)

The following table demonstrates the hardship for low wage earners caused by lack of benefits.

Benefits for Wage and Income Groups

Benefit	Low-Wage/ Income	Mid-Wage/ Income	High-Wage/ Income
Health Coverage with Employer Contribution: Individual Worker And Family	34%	87%	94%
Paid Time Off For Personal Illness	39%	74%	90%
Paid Holidays	46%	86%	89%
Any Retirement Plan with Employer Contribution	32%	72%	87%

(*Understanding Low-Wage Work in the United States*, The Mobility Agenda)

Contingent workers, who are also underemployed, are defined by the Department of Labor as: "Persons who do not expect their jobs to last or who reported that their jobs are temporary." Contingent workers are defined as a worker who fit into one of the following categories:

- on-call workers – workers who are called to work only as needed, although they can be scheduled to work for several days or weeks in a row;
- temporary help agency workers – workers who are paid by a temporary help agency, whether or not their job was temporary; (*Contingent and Alternative Employment Arrangements*, February 2005, Bureau of Labor Statistics)

Contingency workers face similar problems as part-time workers in that a high percentage of contingency workers want and need full-time jobs in order to pay their bills. According to the Department of Labor, 62.7 % of contingent workers, in 2005, prefer full-time jobs. As well, contingency workers accounted for 4.1% of total employment in 2005. (*Contingent and Alternative Work Arrangements*, Bureau of Labor Statistics) Therefore, 2.6% of workers who were defined as employed by the Department of Labor were underemployed since they preferred full-time work. The second step in calculating the percentage of workers who are underemployed is to add 2.6 to the percentage of underemployed workers for a total of 12.6%. (10%+2.6%)

Furthermore, contingency workers earned less money and enjoyed fewer benefits than full-time workers according to the tables above.

Another group of workers who are treated as employed are those who work full time but whose income falls below the poverty threshold. An ongoing problem that prevails in policy discussions about the poor or unemployed is the definition of poverty which is arbitrary and misleading. By using a poverty threshold which doesn't reflect the essential needs of people and which ignores people above the threshold who experience hardships, the Department of Labor is producing a deceptively low statistic about who can meet their basic needs. (Poverty threshold is examined in the next chapter)

According to the Bureau of Labor statistics:

> Of the 145.2 million individuals in the labor force for 27 weeks or more in 2006, 3.7 percent of those usually employed full-time were classified as working poor, compared with 11.3 percent of part-time workers.
> (*A Profile of the Working Poor*, 2006, U.S. Bureau of Labor Statistics)

However, the official poverty threshold is too low and under represents the number of working poor. Using a different measure of poverty which more accurately reflects the number of people living in deprivation yields significantly different results for the percentage of full-time workers who fall below the poverty line.

Rather than using the current one-dimensional poverty threshold which only measures income deprivation, a number of scholars (Shawn Fremstad) advocate a measure of social inclusion which assesses the extent to which low-paid workers lag behind other workers. The current poverty measurement was devised in the 1950s when housing costs were a smaller proportion of household budgets and when child care costs were a minimal factor in monthly expenses.

Social inclusion is defined as an income which not only allows a family to meet its basic needs such as shelter, food, clothing and transportation but also health care, child care, leisure time activities and pension investments. It is also a relative measure in the sense that when incomes in general move upward, the higher the median, the higher the poverty threshold. In other words, as growth trends move upwards, everyone benefits.

A measurement based on social inclusion compares a workers wage to the income dispersion of other jobs to define low-wage work. According to the study *Understanding Low-Wage Work in the United States*:

> To define a low-wage job as one that pays much less than jobs in the middle and upper share of the labor market. The Organization for Economic Cooperation and Development (OECD), with 30 member countries including the United States, and many researchers, already use this method. Accordingly, we define a low-wage job as one that pays less than two-thirds of the median wage for men...to limit the extent to which gender inequality in wages affects the definition of low-wage jobs.
> (*Understanding Low-Wage Work in the United States*, Centre for Economic Policy and Research)

The median wage represents the wage in the middle when arranged in either ascending or descending order. In other words, half the wages are above the median and half the wages are below the median. In order for a person to be considered socially included, they would have to earn at least two-thirds of the median wage.

The median wage is chosen as the measure of choice for defining poverty rather than the mean (average), because the mean tends to be distorted by extreme values at either end and in the case of income, the small percentage whose incomes are in the millions of dollars would result in a much higher mean whereas the median is immune to extreme values and is more representative of the dispersion of wages.

The median wage for all workers in the United States was $16.88 (see table above Re. percentiles) in 2007. Two-thirds of the median wage is $11.25 per hour or $23,400 per year. According to the percentile tables above, 25% of all workers earn below $11.71 an hour which is slightly higher than the social inclusion poverty rate of $11.25 per hour. Heather Boushey, of the Economic Policy Institute, concludes that: "For one out of three working families with young children, income alone is not enough to make ends meet." (*The Needs of the Working Poor*, Heather Boushey) Alan Barber, of the Centre for Economic and Policy Research concludes that: "In the typical state, 22 percent of people in working families suffer from economic hardship." (*Levels of Economic Insecurity* Job *and Quality in the States in the First Half of the 2000s*, Alan Barber)

To be conservative, 22% will be selected as the percentage of the total workforce representing the working poor. Since there is overlap between the percentage of working poor and temporary and part-time workers living in poverty, the latter rates must be subtracted from the percentage of workers who are defined as the working poor.

A problem arises in the above calculation because the percentages for part-time and temporary workers are based on the Department of Labor statistics while the poverty rate for the working poor is based on the social-inclusion approach. To adjust for this inconsistency, a ratio of the two poverty rates will be used to adjust the 22% working poor. In 2006, 35 million workers lived below the poverty threshold whereas 44 million lived in poverty based on the social-inclusion approach. The ratio is .795 (35 divided by 44) and the 22% becomes 17.5% (.795 X 22%). Since there is overlap in the two categories part-time/contingency and those working full-time but living in poverty, it is necessary to subtract 6.8% (4.2% part-time + 2.6% contingency) from the 17.5% for a total of 10.7%. Adding the rate of unemployment already established to 10.7% yields 23.3%.

Surprisingly, when underemployment is used as the measure of the hardships endured by workers, the result is 23.3% rather than 5.8%. The vast gap between the two measurements strongly indicates that workers did not benefit from growth. As well, other factors must be considered to provide an accurate picture. For example, the weakening of the union movement and the ineffectiveness of the market system to distribute incomes fairly are factors as well.

There is one other group who are unemployed but are not represented in the unemployment statistics. There is a group of people who lost their jobs and subsequently searched for work but without any success. These workers often become discouraged and quit seeking employment in the belief that no work existed for them. Workers who have been available for work and looked for a job sometime in the previous 12 months but have not searched in the prior four weeks are referred to as "marginally attached to the labor force". The official definition of unemployment excludes any person who has not been searching for work in the previous four weeks and therefore, discouraged workers have not factored into the unemployment statistics.

In 2008, 1.9% of unemployed workers were classified as "marginally attached to the labor force". (*Ranks of Discouraged Workers*, Bureau of Labor Statistics) The final total for underemployed workers is equal to 23.3% + 1.9% = 25.2%.

Shockingly, more than one in four Americans is defined as underemployed while the official statistics only report 5.8%, a gap that hides the truth about the status of the workforce.

There are a number of assumptions underlying the calculation of the total percentage of underemployment but, nevertheless, it offers a warning that the problem transcends the simple dichotomy of working or not working. To address the problem of underemployment first requires that the statistics underlying the framing of policy must be clearly understood.

Growth clearly has not benefited workers as claimed by those who stake their reputation on an increasing GDP. To make matters worse, workers have actually lost ground since 1970 in terms of real wages or wages adjusted for inflation. Their income today has less purchasing power than 1970 because inflation has outpaced growth in income. According to Eduardo Porter in the New York Times:

> The portion of the economy going to workers in wages and benefits is perhaps the broadest measure of the workers' share of economic growth...In real terms, the wages of nonmanagement employees in the United States are now 10% below their level in the early 1970s, according to Labor Department Statistics.
>
> (*After Years of Growth, What About Workers' Share?*, The New York Times)

The Bureau of Labour Statistics documents the decline in real wages from 1970 to 2004 clearly revealing the decline in purchasing power of workers' income.

Real Wages 1970 to 2004 in 1982 dollars

Year	Average weekly earnings
1970	$312.94
1975	$305.16
1980	$281.27
1985	$276.23

1990	$262.43
1995	$258.43
2000	$275.62
2004	$277.57

(*Wages and Benefits*, Real Wages (1964-2004))

According to this data, wages have declined in real terms from 1970 to 2004 by 11.3% while at the same time, the GDP rose by 61.45%. In other words, not only did workers not benefit by growth but lost a considerable amount of their purchasing power. This discrepancy between growth and the purchasing power of workers demonstrates the futility of interpreting GDP as a measure of well-being. It also reveals that growth in economic wealth is completely irrelevant to the lives of workers.

If growth is so destructive and at the same time not benefiting working people, then it seems pointless to base social and economic policies on the apotheosized GDP. Not only are workers ignored as a result of these policies but so is the environment. A paradigm shift is imperative to construct humane and environmentally social and economic policies.

CHAPTER FOUR

Pay Up:
The Benefits of Growth Exclude the Poor

"A nation's greatness is measured by how it treats its weakest members." (Unknown source) This tenet is singularly true when a country has the resources to eliminate poverty but fails to do so for any one of a number of reasons including misplaced spending, misplaced priorities, or lack of sympathy for those who are vulnerable such as the old, poor, young, disabled or those suffering from a mental or physical disability. America's high growth rate has generated a copious supply of new wealth yet successive U.S. governments have failed to distribute it to those most in need. Invoking the worn out shibboleth about "it's their own fault" fails to recognize that many other countries with less wealth have virtually no poverty.

In fact, America has a higher level of poverty than many countries that are far less wealthy, according to conventional measurements. Compared to 24 OECD countries, the United States ranks last in the percentage of children living in a household where the equivalent income is less than 50% of the national medium. (*Child Poverty in Perspective: an overview of child well-being in rich countries*, UNICEF) That means, in effect, children living in families whose income is at the 25% percentile.

In addition, the Census Bureau claims that 13 million children or 18% live below the federal poverty level ($22,000 for a family of four). The National Centre for Children in Poverty argues that families require twice that income to meet their basic needs, a standard that would mean that 39% of children actually live in low-income families.

Comparing different studies about levels of poverty exposes the fact that calculating a threshold is a daunting task with many variables and parameters. As a result, the threshold chosen can either grossly under-report or over-report the number of people who struggle to survive. Nevertheless, national policies depend on accurate measurements to frame solutions for crises such as poverty.

The American poverty threshold is too low and misrepresents the number of people who are living in poverty. The number of people living in poverty remains high despite the fact that long term growth rates have been steadily rising. Therefore, pressure on the government to implement strong policies to assist the poor are weak since the number of people who need assistance is understated.

Knowing the full extent of poverty is critical because of its deleterious impact on children and in the long run, on society as a whole. Poverty is not just deprivation of basic material necessities such as food, clothes and shelter but also a lack of human and social capital which includes education, basic life skills and employment experience. Without a single-payer, government-run healthcare system, those below the poverty threshold may be unable to afford decent medical diagnosis and treatment.

Child poverty not only robs children of the opportunity to reach their potential in life, it also robs society of people who eventually could participate fully in the economic, political and civil facet of community life. Their mental, physical and emotional development may be at risk for children living in low income families.

Another of the tragedies of poverty is the impact it has on students' ability to succeed in school. Inadequate nutrition not only affects the development of the body, but the brain as well. Malnourished children will have a harder time concentrating and learning. Problems such as crime, absenteeism, low birth weight, child abuse, missing children and attempted suicides are linked to poverty.

As child poverty demonstrates, there is no correlation between growth and the well-being of people in a society. As the basic engine that drives economic and social policies, growth misleads the public into believing that these policies are efficacious when they are not. The determinants of well-being are related to tax policy and who benefits from growth and the direction of social and economic policies. The level of subsidized daycare, subsidized housing, accessibility and affordability of health-care, unemployment insurance and the cost of educations are all critical factors in deciding the degree of poverty in a society. Growth by itself is equivalent to a blank slate and is not automatically beneficial.

A high growth rate in the U.S. has not resulted in effective programs and services for the poor. Despite advocating growth as a means to achieve prosperity which then raises everyone's standard of living and provides jobs, government has instead implemented policies that ignore poverty and underemployment.

In chapter 2, it was demonstrated that in comparison to eight other countries, the U.S. has the highest child poverty rate; the second highest infant mortality rate; and ranks 37th in health care.

Notwithstanding its wealth, education is another area where the United States ranks below other countries with less wealth. Prosperity depends to a large extent on human capital requiring students and adults to advance their knowledge and skills throughout their lives. A heavy dose of the humanities is essential for understanding oneself and the world.

Again, there is no correlation between growth and performance in school. For example, consider the percentage of children enrolled in primary school:

UNICEF – State of the World 2008

Country	Primary School Net Enrolment Percentage
Canada	100
France	99
Germany	98
Netherlands	98
UK	98
Belgium	97
Denmark	96
Sweden	95
U.S.	92

(UNICEF)

In addition to enrolment, the U.S. ranks last in a series of major studies conducted by the OECD concerning the performance of high school students in the areas of mathematics, science and reading. To measure performance in education:

> The OECD launched the OECD Programme for International Student Assessment (PISA) in 1997. PISA represents a commitment by governments to monitor the outcomes of education systems in terms of student achievement on a regular basis and within the internationally agreed common framework.
>
> (Science Competencies for Tomorrow's World, OECD, p. 3)

The results of the science PISA scores are as follows:

PISA Science Scores - 2006

Country	Score
Canada	534

Netherlands	525
Germany	516
UK	515
Belgium	510
Sweden	503
Denmark	496
France	495
U.S.	489
OECD Average	500

(*Mean Scores on the Science Scale*, OECD)

Note that the United States ranks ninth among the nine countries above. Since there is no correlation between growth and the quality of education, factors other than growth are responsible for this poor performance.

The PISA results for mathematics show a similar trend:

PISA Mathematics Scores - 2006

Country	Score
Netherlands	531
Canada	527
Belgium	520
Denmark	513
Germany	504
Sweden	502
France	496
U.K.	495
U.S.	474
OECD Average	498

(*Mean Scores on the Mathematics Scale*, OECD)

Again, note that the U.S. ranks last.

High rates of poverty cause surprisingly high costs to society in the form of health care, crime, lost productivity and lost consumption. According to the National Centre for Children in Poverty, child poverty costs the U.S. $500 billion per year. (*Ten Important Questions about Child Poverty and Family Economic Hardship*, National Centre for Children in Poverty)

By defining a poverty threshold that excludes so many people suffering from hardship, the government is hiding a serious problem that is costly in terms of human adversity and loss of human capital.

To understand the process for constructing a threshold that includes all families experiencing deprivation, and at the same time, excludes families that are not, it is necessary to examine a number of principles and parameters involved in measuring poverty.

One of the first parameters to consider is whether the threshold will be a relative or absolute measure. An absolute measure is usually based on the cost of resources needed to maintain a minimal standard of living. Absolute measures differ as to the items included in the calculations but all absolute measures would at least incorporate the cost of shelter, food and clothing.

Relative measures are related to the standard of living of society so that when the standard of living increases so does the poverty threshold. The theory underlying relative measures is the belief that deprivations are to be assessed according to the well-being of others. In other words, the poverty threshold is relative to living standards in society.

The United States currently uses an absolute measure based on the cost of the Department of Agriculture's "thrifty food plan", developed by economist Mollie Orshansky, adjusting it for family size and composition, and then multiplying it by three. The factor three was adopted because in the 1960s, it was believed that a family spent one third of its budget on food. In addition, the threshold was adjusted each year based on the consumer price index.

As per the discussion in Chapter 2, the revision of the CPI in the 1990s understates the real increase to the cost of living. Applying the new CPI to the poverty threshold does not increase it sufficiently to reflect the real change in the number of people living in adversity.

Revisions to the CPI are only one of many criticisms of the current measure of poverty. Another problem is the assumption that families spend one third of their income on food. This assumption is not valid today. This reflects a more general problem regarding the relative weightings in the index of goods and services. Family consumption habits change overtime and should not be fixed based on weightings over 40 years old.

Additionally, expenses that were not relevant in the 1960s are highly relevant today. Examples include child care expenses, now common due to the increased number of women who have entered the labor force. A number of expenses are excluded that might be considered essential for a family to avoid suffering hardship. These include child care expenses, health care expenses, transportation, educational expenses (example: field trips) and saving for retirement when it is not a work benefit.

There are two measures of poverty in the United States. First, poverty thresholds are issued in a report produced by the U.S. Census Bureau based on the method discussed above and show how poverty is distributed by age, race, region and family type. Second, poverty guidelines are issued each year by the department of Health and Human Services and are a simplification of the poverty thresholds to determine eligibility for federal programs.

Census Bureau's poverty thresholds for 2006

Size of Family Unit	Poverty Threshold
One person	$10,294
Two people	$13,187
Three people	$16,227
Four people	$20,444
Five people	$24,059

(*Poverty thresholds 2006,*U.S Census Bureau)

To demonstrate the inadequacy of this threshold, it will be compared to poverty thresholds used in Canada. Canada uses a relative measure called Low Income Cut-offs (LICO) although Statistics Canada, who calculate the LICOs, avoids the term poverty line or poverty threshold but defines them as indicators of the extent to which some Canadians are less well-off than others based solely on income. The intent of the cut-offs is to set a threshold below which a family is spending too much of its income on food, shelter and clothing.

Calculation of the LICOs is based on the 1992 Family Expenditure Survey and the incomes of 9,472 private households across Canada. The data was divided into seven household sizes and cross-referenced with five community sizes. Then for each category, Statistics Canada calculates the average percentage that a family spends on food, shelter and clothing. Statistics Canada assumes that a family spending more than 20% of this average on the three basics is a low-income family. A family with an income below the cutoffs would not have sufficient income for other necessities such as dental care, school expenses and transportation.

To determine the amount an average family spends on the three basics, Statistics Canada chose 1992 as the base year. In 1992, the average family spent 43% of their before tax income on these basic necessities. Therefore, if a family spends more than 63% of its income on food, shelter and clothing, they are considered to be low income.

Canadian LICOs for the year 2006 are as follows:

<u>Low income before tax cut-offs, 2006</u>
Size of Area
Of Residence

Family Size	100,000 to 499,999	500,000 or More
1	18,257	21,119
2	22,728	26,392
3	27,941	32,446
4	33,925	39,393
5	38,476	44,679

(*Low income cut-offs before taxes*, Statistics Canada, p. 24)

There are a number of problems with the LICOs. Statistics Canada has decided not to repeat the 1992 survey so that current spending patterns are not reflected in the LICOs, rather they are fixed at 1992 levels. The only adjustment to the LICOs is the cost of living based on the Consumer Price Index. Secondly, the 20% is arbitrary and does not necessarily reflect the additional burden entailed in maintaining a standard of living commensurate with the rest of society. The survey on which the cutoffs are based may not accurately reflect expenditures on the three necessities.

Canada had developed another relative low income measurement, Low Income Measures (LIM), primarily for the purpose of international comparisons. Whereas the LICOs are expenditure based, the LIMs are income based. It is a truer relative measure of poverty because it is based on median income rather than on a fixed measurement of the cost of basic necessities.

LIMs are determined by calculating 50% of the median of family income adjusted for size. Adjusting for size involves adding 1 for the first person in the family and .4 for the second, regardless of age. Additional adults are counted as .4 and each child is counted as .3. Family income is then divided by the total adjustment factor to arrive at the adjusted family income. For example, if a family of four with two children and two adults has a pretax income of $60,000, the adjusted income is calculated by dividing 60,000 by two (1+.4+.3+.3) or $30,000. A family with an income of $60,000 with two adults and three children would have an adjusted income of $26,091. When calculating the median, the second family's adjusted income is lower due to its larger size. The LIM for a family of a particular size is 50% of the median based on adjusted family incomes.

Canadian LIMs for the year 2006 are as follows:

Low income measures before tax 2006

Number of adults	Number of Children			
	1	2	3	4
2 adults	29,643	34,874	40,105	45.336

(*Low Income Cut-offs for 2007 and low income measures for 2006*, Statistics Canada)

Observe that the LICO for a family of four (in a large urban centre) is $39,393 while the LIM for a family of four is $34,874.

There are other measures of poverty including the Market Basket Measure demonstrating that low income measures can vary considerably depending on which method is used. To address the crisis of poverty requires a reasonably accurate measure so that politicians must choose carefully when attempting to assess poverty.

Before comparing the poverty threshold in the United States to the thresholds in Canada, it is necessary to convert the Canadian dollars to American dollars.

The following table compares the two Canadian thresholds to the American threshold for the year 2006 after converting Canadian dollars to American dollars based on the exchange rate on June, 30, 2009:

Comparing U.S. and Canadian thresholds 2006 ($USD)

Method	LICO*	LIM	U.S. Threshold
Number of Family Members			
1	18,944	26,590	10,294
2	23,674	31,282	13,187
3	29,104	35,974	16,227
4	35,335	40,666	24,059

(See above tables)

*Assuming families living in urban centre of 500,000 plus

The U.S. thresholds are well below the LIMs and LICOs in Canada which raises the question about whether the U.S. poverty numbers are too low thus leading policy makers and the public to wrong conclusions about the severity of the problem.

Furthermore, the OECD, European Union, and the United Nations use 50% of median income to construct a poverty line: a relative measure that states that if a family earns less than half of the medium for family income, they are considered low income.

Applying the same measure as the above organizations to the United States would yield a higher number of people in the United States who live in poverty. Using the year 2006 as an example, the median income for a 4-person family is $67,019. (Median Income for 4-person Families, by State, U.S. Census Bureau) Therefore, the poverty threshold based on 50% of the median would be $33,570. Note that this poverty threshold is almost identical to the one arrived at in Canada using LICO's. Also note how this measure of poverty is so much higher than the one currently in use in the United States ($24,059).

Using 50% of the medium results in a higher poverty rate and comparing the American rate to the same eight other OECD countries used in the past produce the following result:

<u>Poverty rate after transfers and taxes – poverty mid-2000s</u>

<u>Country</u>	<u>Poverty rate</u>
Denmark	5.3
Sweden	5.3
France	7.1
Netherlands	7.7
UK	8.3
Belgium	8.8
Germany	11
Canada	12
U.S.	17.1

(*Income distribution poverty*, OECD)

Although direct comparison is difficult because the OECD data is based on after tax and transfers income, it is safe to say that the poverty rate is much higher using 50% of the median as the threshold. Observe that the U.S., despite being the wealthiest country in the world, yet again ranks last and that five of the above countries have a poverty rate that is half that of the U.S.

Poverty is only one aspect of a larger problem: income inequality. Growth has benefited very few people and has left behind far too many people to suffer declining family incomes, unemployment, underemployment and loss of social services resulting in a lower standard of living.

One method used to establish the inability of growth to spread prosperity is to again compare inequality of income in the wealthiest country in the world with other wealthy countries.

A common technique for such a comparison is to use the Gini coefficient, developed by statistician Corrado Gini in 1912, which is commonly used as a measure of income inequality. The Gini coefficient can range from 0 to 1 with a coefficient of 0 indicating perfect equality of income and 1 representing perfect inequality. Perfect inequality would be a society in which one person receives 100% of all income.

Comparing the Gini Coefficient to the eight OECD countries reveals the following:

Income Inequality mid-2000s

Country	Gini Coefficient
Denmark	.23
Sweden	.23
Netherlands	.27
Belgium	.27
France	.28
Germany	.30
Canada	.32
UK	.34
U.S.	.38

(*Measures of Income Inequality*, OECD)

As a reminder of the wealth of these eight countries measured in GDP per capita, consider the following table:

GDP Per Capita 2008 (est.)

COUNTRY	PER CAPITA GDP
U.S.	$47,000
Netherlands	$40,300
Canada	$39,300
Sweden	$38,500
Belgium	$37,500
Denmark	$37,400
UK	$36,600
Germany	$34,800

France $32,700

(*Country Comparison – GDP-per capita*, Central Intelligence Agency <CIA> Fact Book)

It is abundantly clear that there is a complete lack of any correlation between growth and a fair distribution of income. The U.S. ranks 27th overall among the 30 OECD countries in income inequality.

Another method for measuring income inequality is the decile ratio which compares how much greater the income of the people at the 90% percentile is to the income of the people at the 10% percentile. In other words, how much more to the richest people earn compared to the poorest. According to this measure:

Income inequality mid-2000s

Country	Decile ratio
Denmark	2.72
Sweden	2.79
Netherlands	3.23
France	3.39
Belgium	3.98
Germany	3.98
Canada	4.12
UK	4.21
U.S.	5.91

(*Measures of Income Inequality*, OECD)

The U.S. has by far the most unequal distribution of wealth and ranks 29th overall among the 30 OECD countries. In the United States, the income at the 90th percentile is 5.91 times greater than the income at the 10th percentile.

Confirming previous conclusions, the data on poverty irrefutably demonstrates that there is no correlation between growth and quality of life, according to standard of living criteria. Other policies and indicators are needed to address the problems of unemployment, underemployment, poverty and the environment. The problem is much deeper than indicators and policies and is related to an underlying ideology that governments have adopted to guide policymaking. Neoliberalism and free markets are at the core of this ideology which minimizes the government's responsibility for redistributing wealth. Markets are not really free and are stacked in favor of wealthy individuals and corporations. In addition to people being victimized by free markets, the environment is virtually neglected.

Voluntarism has not solved these problems in the past and there is no reason to believe that it will in the future. For example, poverty will not disappear until the government implements policies which reverses the flow of benefits of sustainable growth so that those who are deprived can enjoy some of the benefits as well. In a similar vein, only the government can ensure that growth is sustainable.

Fear of government must be dispelled and people must understand that it is the people running the government who must be held accountable. Lincoln's insightful words "government of the people, by the people, for the people" imply that an attack on the institutions of government is really an attack on the people. An effective functioning government would serve the interests of the people.

CHAPTER FIVE

Drink Up:
Shrinking the Supply of Potable Water

In past years, depletion of resources usually conjured up images of idle oil wells or vacant mines. Only in recent years have people begun to regard water and air as finite resources with limited capacity to accommodate pollution. At a certain point, contamination of air and water will cause serious damage to humans and the ecosystem. The difference between running out of oil and running out of potable water is that the former can be replaced with other resources but the latter can not be replaced despite the burgeoning desalination industry.

Although the prospect of depleting all the fresh water resources on the planet seems remote, according to the U.S. General Accounting Office (GAO):

> The nation's capacity for storing surface-water is limited and groundwater is being depleted. State water managers expect freshwater shortages in the near future, and the consequences may be severe. Even under normal conditions, water managers in 36 states anticipate shortages in localities, regions, or statewide in the next ten years.
> (*States' Views of How Federal Agencies Could Help Them Meet the Challenges of Expected Shortages,* General Accounting Office)

In reality, the growing water shortage is invisible in a society that measures progress in terms of unsustainable growth. Measuring progress without incorporating all the costs is simply postponing or transferring the problem to future generations.

Make no mistake; environmental crises such as the growing shortage of water cannot be viewed in isolation due to the complexity and interactions among various systems in the ecosystem. Weather systems, carbon cycle, hydrologic cycle, biodiversity and living organisms have complex interdependent and synergetic relationships which defy simple analysis.

For example, deforestation and fossil fuels contribute to global warming. This in turn affects the availability of fresh water in a number of ways including melting of glaciers, evaporation, irrigation, and desertification. Shockingly, according to the World Health Organization, the growing shortage of clean water is responsible for 80% of sickness and disease worldwide. (*Blue Covenant: The Global Water Crisis and the Coming Battle for the Right to Water*, Maude Barlow, p.3) Because of this tragedy, high infant mortality rates in poor countries encourage families to have more children and the expanding population makes greater demands on water supplies.

Further, the loss of forests, clean water and clean air is partly due to our continuing use of fossil fuels as a source of power. The United States and Canada are lagging behind other countries in switching to alternative forms of energy such as solar, geothermal tidal and wind power. Currently, only 7% of U.S. energy sources are based on alternative technologies as shown in the following table:

Energy Consumption 2008

Source	Percentage of total Energy Supply
Petroleum	37%
Nuclear	9%

Natural Gas   24%
Coal   23%
Renewable   7%
(*Renewable Energy Consumption and Electricity Preliminary Statistics 2008*, U.S. Department of Energy)

A breakdown of the renewable energy sources reveals the following:

Renewable Energy Consumption

Source	Percentage Of Total
Solar	1%
Hydroelectric	34%
Geothermal	5%
Biomass	53%
Wind	7%

(*Renewable Energy Corporation and Electricity Preliminary Statistics 2008*, U.S. Department of Energy)

The primary source of renewable energy is biomass which constitutes over half of all energy produced by sustainable means. An OECD study on the use of biomass or biofuels reports that:

> The high level of policy support [for biofuels] contributes little to reduce greenhouse-gas emissions and other policy objectives, while it adds to a range of factors that raise international prices for food commodities... The report concludes that there are alternatives to current support policies for biofuels that would more effectively allow governments to achieve these objectives.
>
> (*Biofuel Support Policies*: An Economic Assessment, OECD)

According to the report, biofuel production has limited impact on reducing greenhouse gasses and is highly dependent on government funding. Also, using wheat, sugar, beet, vegetable oil and, in particular, corn for biomass fuel increases demand for these products. The result is a rise in the average wheat prices by 5%, maize by approximately 7%

61

and vegetable oil by 19% over the next ten years. These increases in basic food products will have a devastating effect on the poor in the United States and around the world.

Another problem is that the United States, the largest consumer of energy worldwide, lags behind other countries in the development of renewable sources of energy. Despite reports that claim that the United States is the leader in developing renewables, a closer examination of these studies reveals that faulty methods were used in calculating the numbers used for the comparison. Absolute numbers such as megawatts were used while ignoring a number of problems with this approach. The absolute numbers should have been compared to GDP or population for standardization purposes which allowed for the wealth or size of the country.

For example, the United States produces 25,170 Megawatts of electricity using wind power compared to Germany which produces 23,900. Adjusting the numbers for population (The U.S. has 3.7 times more people), the figure for the United States would be 6803 megawatts. An important consideration is the total amount of power produced by each country and the fraction represented by renewables. The following table provides some examples of the ratio of renewables to the percentage from all sources:

Country	Renewables as a percentage of total energy produced - 2008
Germany	15%
Denmark	15%
France	12%
U.S.	7%
UK	2.1%

Further, when you consider the top ten countries with the greatest wind power capacity, the results are completely different after you adjust the number of megawatts either for population or for GDP as shown in the following table.

Ranking Wind Production - 2008

Country	Output in Megawatts (MW) Before adjustment	Ranking out of top 10 countries	Output (MV) Adjusting for Population	Ranking
U.S.	25,170	1	.084	4
Germany	23,900	2	.290	3
Spain	16,740	3	.410	2
Italy	3,740	6	.064	5
France	3,400	7	.053	6/7
UK	3,240	8	.053	6/7
Denmark	3,180	9	.580	1

(International Monetary Fund for GDP)

(*Renewable Global Status Report*, International Energy Agency)

Ranking countries according to megawatts per capita seems to be the most sensible method for standardization in which case, the United States ranks fourth among the countries studied above. It also ranks fourth in terms of the percentage of energy derived from renewable sources among the five countries above.

Despite large new investments in wind power and other renewables, it is evident that the United States is far short of the mark in transferring to alternative forms of energy. America is not alone in lagging behind in timetables suggested by scientists. Fossil fuels must be phased out within the next 30 or 40 years in order to control the damage it causes. Scientists warn that we must reduce greenhouse emissions from between 80% and 90% by 2050 requiring the government to adopt strong measures to reach that target. There is still much resistance from the oil companies and only strong government action will achieve a viable result.

One of the more critical losses resulting in part, from our use of unsustainable energy is the imperiled state of the world's water supply. Indirectly, fossil fuels impact the world's availability of clean water.

It works as follows: fossil fuels contribute 80% of the carbon dioxide in greenhouse emissions and furthermore, carbon dioxide represents 85.4% of total greenhouse emissions. (*2009 U.S. Greenhouse Gas Inventory Report*, EPA) Fossil fuels contributed 68% of all greenhouse emissions which contribute to global warming and therefore, impact the water supply. For example, climate change is melting glaciers which are essential in providing clean water for hundreds of millions of people.

The Colorado River demonstrates the issue of declining clean water levels. Consider the seven States (Arizona, New Mexico, Utah, Colorado, Wyoming, California and Nevada) which depend on the Colorado River Basin for their drinking water, hydro electricity and irrigation. The Colorado Basin provides drinking water for approximately 30 million people and irrigates 3.5 million acres of farmland. Lake Powell alone generates sufficient electricity to serve 1.5 million homes.

Recent studies have shown that population increases and climate change have been significant factors in the growing gap between supply and demand for water from the Colorado River Basin. According to a National Academies of Sciences' report:

> Recent studies of past climate and streamflow conditions have broadened understanding of long- term water availability in the Colorado River, revealing many periods when streamflow was lower than at any time in the past 100 years of recorded flows. That information along with two important trends-a rapid increase in urban population in the West and significant climate warming in the region-will require that water managers prepare for possible reductions in water supplies.
> (*Colorado River Basin Water Management: Evaluating and Adjusting to Hydroclimatic Variability*, National Academy of Sciences, p.1)

According to a major study published in Water Resources Research in 2006:

> Severe drought conditions in the Colorado River basin, coupled with a large increase in water use over the past two decades, have recently resulted in water demands that have outstripped natural inflows.
> (*Updated Streamflow Reconstructions for the Upper Colorado River Basin*, Woodhouse C. A., Gray S. T. and Meko D. M., p. 1)

Confirming the impact of global warming on the water supply, the same study reports that:

> Demands on the Colorado River over the past decades have risen to meet or exceed average availability. Any variations or shifts in climate can have a significant impact on the system.
> (ibid; p. 15)

On a global level, shortages of clean water have had a tragic impact. Consider the following facts:

- Water scarcity may limit food production and supply, putting pressure on food prices and increasing countries' dependence on food imports.
- More children are killed by dirty water than by war, malaria, HIV/AIDS and traffic accidents combined.
- Two billion people live in water-stressed regions of the planet.
- Parts of the world that are running out of potable water include Northern China, large areas of Asia and Africa, the Middle East, Australia, the Midwestern United States and sections of South America and Mexico.
- Two-fifths of the world's people lack access to proper sanitation, which has led to massive outbreaks of waterborne diseases.
- Annual rainfall is declining; salinity and desertification are spreading rapidly and rivers are being drained at an unsustainable rate.

- In 2007, Lake Superior, the world's largest fresh water lake, dropped to its lowest level in 80 years and the water has receded more than 15 meters from the shoreline.
- Ninety percent of wastewater produced in the Third World is discharged into local rivers, streams and coast waters.
- In China, 80 percent of the major rivers are so degraded that they no longer support aquatic life.
- In Pakistan, less than 25 percent of the population has access to clean drinking water due to massive pollution of the country's surface waters.
- Seventy-five percent of India's rivers and lakes are so polluted they would not be used for drinking or bathing in North America.
- In India, 2.1 million children under the age of five die each year from dirty water.
- Seventy-five percent of Russia's inland surface water is polluted and thirty percent of the groundwater available for use is highly polluted.
- Twenty percent of all surface water in Europe is seriously threatened.
- Forty percent of U.S. rivers and streams are too dangerous for fishing, swimming or drinking.
- In Latin America and the Caribbean, more than 130 million people do not have safe drinking water.
- More than one-third of African's population currently lacks access to safe drinking water.
  (*Blue Covenant: The Global Water Crisis and the Coming Battle for the Right to Water*, Maude Barlow)

Clearly, unsustainable growth has had a devastating effect on our water supply leaving us with a dwindling supply of clean water to meet our basic needs.

Desalination plants are being constructed as a magic bullet to solve the problem of water shortages. In the process, past patterns are being repeated in that the unsustainable aspects of desalination are overlooked in the rush to apply the new technologies which have been developed to convert seawater or brackish water (water with less salinity than ocean water) to drinking water. Costs, greenhouse emissions, ocean salinity, use of nuclear power, impact on marine life, thermal pollution, diversion from alternatives and corporate control over water are dangers that must be considered and addressed before desalination plants become so numerous that the problem is completely out of control.

The waste water remaining after desalination is a concern because it now has a heavy concentration of saline in addition to other chemicals which are returned back to its original source.

According to a World Wildlife Fund (WWF) study:

> Anything in the source waters can be expected to show up in more concentrated form in the discharges from water manufacturing plants, along with the chemicals added during the treatment processes or from other processes such as corrosion. There are also many thermal issues with the discharges. In the case of seawater desalination, the main discharge issues can include elevated levels of salt and other constituents of seawater such as boron, dead sea life which consumes oxygen while decomposing, chemicals added to change the composition of the water for processing and to reduce contamination and clogging of filter membranes, corrosion byproducts and the heat added during processing.
>
> (Making Water Desalination: option or distraction for a thirsty world, WWF)

The discharge consisting of concentrated seawater can harm ecosystems such as marine environments in certain regions with low turbidity and high evaporation rates where the level of salinity is already high. For example, in the Persian Gulf, Red Sea and coral lagoons of atolls, the ecosystems are endangered as a result of high salinity. Science needs to learn much more about the salinity of the oceans and the impact of it rapidly increasing from desalination plants before pumping more saline and other chemicals back into the ocean.

Another problem with desalination plants is that the intake structures kill fish and other organisms that cannot free themselves from the intake mechanism. The Environmental Protection Agency (EPA) enacted a regulation on July 9, 2004, to minimize the danger of entrainment (drawn into the water with the water source) or impingement (aquatic organisms collide with the intake screens). According to this regulation:

> Today's final rule establishes performance standards that are projected to reduce mortality by 80 to 95 percent and, if applicable, entrainment by 60 to 90 percent. With the implementation of today's final rule, EPA intends to minimize the adverse environmental impact of cooling water intake structures by reducing the number of aquatic organisms lost as a result of water withdrawals associated with these structures.

(National Pollutant Elimination System, Environmental Protection Agency, p. 1)

Despite this regulation, the problem of entrainment and impingement still persist as demonstrated by the desalination plants on the coast of California. Forty-five percent of California's generating capacity using this cooling technology is located along the coast and estuaries and withdraws approximately 17 billion gallons of seawater per day. (*Understanding Entrainment at Coastal Power Plants: Informing a Program to Study Impact and their Reduction*, California Energy Commission) According to a study commissioned by the California Energy Commission in 2007:

> Only seven of the 21 OTC [once-through cooling] plants in California have conducted studies of entrainment effects that meet current scientific standards. All were conducted after 2000, including Encina, Huntington Beach, Morro Bay, Moss Landing, Potrero, San Onofre Nuclear Generating Station (SONGS), and South Bay. Six of these studies at SONGS and Potrero were incomplete with regard to entraiment impacts. Since the publication of Foster (2005), Encina has conducted impact analyses, and SONGS and Potrero have completed ongoing analyses. The remaining studies were deemed inadequate in Foster's (2005) review with regard to entrainment impacts. Therefore, seven studies of California power plants

could be used to determine losses attributable to entrainment. (*Understanding Entrainment at Coastal Power Plants: Informing a Program to Study Impact and Their Reduction,* California Energy Commission)

The dangers posed by desalination plants can be minimized if the government sets strict standards and enforces them. Trusting industry to regulate and monitor itself has failed in most industries in the past. For example, automobile manufacturers have not accepted the responsibility to phase out fossil fuel cars and the pharmaceutical industry charges exorbitant prices during its twenty year monopoly granted to them by intellectual property rules. Twenty years is well beyond the period needed to recover research and development costs. As well, chemical industries have introduced toxic chemicals into the ecosystem. Western oil companies operate in other nations such as Nigeria and Ecuador while collaborating with local dictatorships to destroy their land and deprive local people of the means to survive. The United States seems to be repeating the same pattern of laissez-faire in the desalination industry.

Another problem in the desalination of water is the cost of energy needed to power the plants and the resulting greenhouse emissions. Costs vary considerably depending on the location of the desalination plant and the method used. For example, costs will be very high in developing countries where the water must be piped into the interior or through high elevations.

Prices of desalinated water are coming down but part of the reason is the use of nuclear power plants to furnish the power for these plants. There is support for nuclear power from a number of sources including the International Energy Agency, the American Nuclear Society and the European Nuclear Society.

Despite this support, nuclear power plants are plagued with many problems that remain unresolved. Nuclear accidents are a serious risk that could potentially cause extensive damage. Consider the following accidents:

- March 28, 1979 at Three Mile Island, Dauphin County, Pennsylvania: A partial core meltdown releasing 13 million curries of radioactive noble gases and 20 curries of radioactive iodine.
- July 16, 1979 at Church Rock, New Mexico: Evaporating pond failed; the incident drained 100 million gallons which was discharged into the Rio Puerco.
- September 29, 1979 at American Atomics, Arizona: tritium leak resulted in $300,000 worth of food contaminated at the public school across the street.
- July 1981 at Nine Mile Point, New York: An overloaded waste-water tank was flushed into the waste building sub-basement causing 150 55-gallon drums to overturn discharging contaminated water into Lake Ontario.
- 1982 at International Nutronics, Dover, New Jersey: An unknown quantity of radioactive cobalt was spilled into the Dover sewer system.
- November 3, 2005 in Hadem Connecticut: Water containing quantities of radioactive materials leaked from a spent fuel rod.
- May 5, 2006 at the Prairie Island Nuclear Power Plant in Minnesota: An accidental release of radioactive iodine gas exposed one hundred plant workers to low-level radiation.

Disposal of waste radioactive material from nuclear power plants is a very controversial issue which needs to be resolved urgently as waste continues to accumulate without a safe method for disposal. One of the problems with disposal of radioactive waste is that some of the radioactive isotopes in the waste have a half life of hundreds of thousands of years.

Currently, radioactive waste is stored on site in steel-lined, concrete pools or basins filled with water or in massive, airtight steel or concrete canisters. Yucca Mountain in Nevada has been under consideration as a disposal site since 1987 as a centralized, deep geological repository.

In 2002, Congress approved a plan to dispose of waste in Yucca Mountain but President Obama has rejected the plan. The problem now is that there is no plan for disposal and the on-site storage facilities are almost at capacity. According to study undertaken by the Institute for Energy and Environmental research:

> Corrosion of the metal canisters has been a critical problem in assessing the suitability of Yucca Mountain as a deep geologic repository. While DOE [Department of Energy] believes that certain corrosion problems are insignificant, other researchers have concluded that the problem is fatal to the DOE's design of an unsaturated repository – that is a repository above the water that has water vapor and air in the rock pores.
> (*Nuclear Power and CO2 Emission reductions Comments on Radioactive Waste Management and Relative Costs of Options*, Arjun Makhijani)

Waste disposal is a problem that has been ignored for too long and current methods are unsustainable. The cost of safe disposal has been externalized and now we will have to pay the price.

A danger of desalination that is not as obvious is the transfer of the control over water from the public sector to the private sector. Water is a basic need and should be universally accessible and affordable and only the government will observe these principles. Once the private sector gains control over water supplies costs will rise, service will deteriorate and quality will suffer because of the profit imperative.

As of 2007, there were already approximately 87 desalination plants in the world. Suez, Veolia, Siemens, Dow, Nalco and GE are the leading corporations who are providing water to thirsty populations.

The following example can best illustrate the dangers of privatizing water utilities: When the Bolivian government was heading into an economic meltdown, it turned to the World Bank for a loan. Bolivia was forced to privatize railways, the telephone system, national airlines and the hydrocarbon industry. When the loan was up for renewal, the World Bank insisted that Bolivia privatize its water services.

Aguas del Tunari, owned by the Bechtel corporation, won the uncontested contract to provide water services to the town of Cochabamba with a promise to expand water services to people who lacked access to clean water and to improve the quality of water. In return, the company was guaranteed a 15% profit annually.

To finance its operations in Bolivia and earn its guaranteed profit, Aguas del Tunari imposed a large increase in the cost of water. Many of the people in the town were forced to pay 25% of their income simply to cover the cost of water.

Setting an example for those who oppose corporate power, the people of Cochabamba fought against the privatization of their water and eventually forced the company to withdraw. It is clear that the privatization of water will enrich corporations but at the expense of the people to whom they are supplying the water.

This example is very instructive because it exposes the dangers of unsustainable growth. Supposedly, the Bolivian economy grew due to the economic activities of a new corporation but, in fact, the profits were channeled back to Bechtel in the United States. However much the GDP would have risen due to the privatization of water, the public were actually far worse off compared to the government-run utility.

In Cochabamba, corporations and profits benefited while the needs of the people in Bolivia suffered. Even deprivation of people's basic right to water did not deter a major corporation from attempting to increase profits through expansion.

CHAPTER SIX

Cut Up:
Deforestation and its Ramifications

*A*t the rate we are destroying forests, the old expression "You can't see the forest for the trees" will soon be replaced with "You can't see the forest **or** the trees". Forests are a complex mechanism in the earth's ecology that interact with many other systems such as the hydrologic cycle, biodiversity, weather, aridity of soil and global temperatures. We are rolling the dice by destroying forests without sustainable steward-ship and a sound reforestation program. Destruction of our forests will have ramifications well beyond losing trees and a carbon sink.

When it comes to forests, unsustainable growth has fostered short-term thinking at the expense of the health of the planet and its ability to sustain life. Economic growth results from the sale of commodities that are produced as a result of cutting down trees for lumber, furniture and paper products. It also results from farm products produced by farms and cattle ranches in the rainforest. Profits from these commodities and the loss of rain forest due to peasants who have to cut down trees for land on which to live and farm, contribute to unsustainable growth.

Unsustainable growth means that costs pertaining to the damage caused by deforestation are not factored into economic decision-making because they are hidden and will not occur for years. Because this is

73

not right before our eyes, they are ignored and growth continues while invisible costs mount.

Poor stewardship of forests is motivated by the huge profit to be earned from forest land or products. Sustainability attempts are meager since most of the new trees are planted to be cut down again for so-called "productive" use. The shortsightedness and foolishness of our management of forest resources is that no price is attached to the damage caused by deforestation making it appear as if the profits from forests have a free ride. Conversely, chopsticks and hamburgers made from beef raised in the rainforest have a known cost but the cost of the ice caps melting or desertification is invisible, at least in the short term.

Forests cover 30% of the total land mass of the planet which is equivalent to 4 billion hectares (ha) (one hectare is equal to 10,000 square meters). World trade in forest products totals $270 billion annually and more than 1.6 billion people depend on forests for their livelihood. (United Nations Environment Program [UNEP])

Deforestation is occurring at an alarming rate with about 13 million ha per year or 365 million square meters a day lost to agriculture, unsustainable harvesting of timber, the creation of human settlements and drilling for oil. Only 20% of forests today are primary forests where there are no clear visible indications of human activity and where ecological processes remain intact. (UNEP)

It is one of the key factors contributing to climate change due to the loss of an important carbon sink. Tropical forests maintain the carbon balance in the atmosphere through the storage of carbon and the release of oxygen back into the atmosphere by respiration. Burning of trees in forests also releases the carbon stored in the trees. UNEP claims that:

> Another important part of the CO_2 in the atmosphere comes from changes in land use, responsible for almost 20% of atmospheric carbon. Trees and other plants remove carbon from the atmosphere in the process of growing. When they decay or are burnt, much of this stored carbon

escapes back into the atmosphere…Deforestation also causes the release of the carbon stored in the soil.
(Kick the Habit: A UN Guide to Climate Neutrality, UNEP, p. 42)

The consequences of global warming are already disastrous. For example, tornadoes, hurricanes, ice storms and floods kill and displace people and cause serious damage to property. Between 1991 and 2005, 3.47 billion people were affected by disasters, 960,000 people died and economic losses totaled totaled $1,193 billion (US). Exposure to weather hazards are increasingly resulting in even greater losses. (*Bulletin: World Climate Conference-3*, World Meteorological Organization [WMO])

Additionally, sea levels will rise between 18 cm and 59 cm which will lead to the flooding of coastal cities worldwide. Death, property damage and displacement of people will be on a massive scale. It is highly probable that typhoons and hurricanes will become more intense with higher wind speeds again leading to death, displacement and property loss. (*Bulletin: World Climate Conference-3*, World Meteorological Organization)

As well, global warming is responsible for water levels dropping in some major rivers including the Yellow River in China, the Ganges in India, the Niger in West Africa, and the Colorado in the U.S. According to Atmospheric Corporation for Atmospheric Research:

> Rivers in some of the world's most populous regions are losing water… that in many cases the reduced flows are associated with climate change. The process could potentially threaten future supplies of food and water… Several of the rivers channeling less water serve large populations.
> (*Water Levels Dropping in Some Major Rivers as Global Climate Changes*, University Corporation for Atmospheric Research, p. 1)

Another disaster to which climate change contributes is desertification affecting 20,000 square miles of land worldwide annually. Desertification occurs in dry land areas where the earth is particularly vulnerable making these areas susceptible to the impact of climate change.

At least 250 million people are directly impacted by desertification and one billion people in more than 100 countries are at risk of suffering the consequences. (*Climate Information for Securing Food*, World Meteorological Organization) Also devastating is the economic loss of US$42 billion in agricultural goods and the resulting increase in the price of food worldwide. (*Climate and Land Degradation*, World Meteorological Organization, p. 8) Other consequences of desertification include famine, decline in the quantity of fresh water, increased poverty and instability.

According to the World Meteorological Organization, climatic factors contribute to land degradation because:

> Climate exerts a strong influence over dry land vegetation type, biomass and diversity. Precipitation and temperature determine the potential distribution of terrestrial vegetation and constitute the principal factors in the genesis and evolution of the soil...The generally high temperatures and low precipitation in the dry lands lead to poor organic matter production and rapid oxidation. Low organic matter leads to poor aggregation and low aggregate stability leading to high potential for wind and water erosion.
> (*Climate and Land Degradation*, World Meteorological
> Organization, p. 11)

The most important climatic factor that contributes to desertification is rainfall. According to the WMO:

> Rainfall plays a vital role in the development and distribution of plant life, but the variability and extremes of rainfall can lead to soil erosion and land degradation. The interaction of human activity on the distribution of vegetation through management practices and seemingly benign rainfall events can make land more vulnerable to degradation. These vulnerabilities become more acute when the prospect of climate change

is introduced.
(*Climate and Land Degradation*, World Meteorological
Organization, p. 13)

Deforesting an area entails destroying the habitat of a number of species. Again, the loss of biodiversity is a crises arising from short-term thinking, greed for immediate profits and growth at the expense of long-term risks. These risks are not assigned a value meaning that their significance is not factored into the cost assessment of deforestation but rather are externalized so that a future generation will be forced to pay the costs.

To understand biodiversity, it is important to recognize that we are part of a complex web of life forms. Disrupting part of the labyrinthian interactions among different species might seem remote and irrelevant but, in fact, it is vital to preserving the ecosystem responsible for creating the conditions of life. According to the Secretariat of the UN Convention on Biodiversity (CBD):

> Although scientists are now able to appreciate the complexity of this web of interacting processes, we are still a very long way from understanding how they all fit together. What we do know is that if any part of the web suffers break downs, the future of life on the planet will be at risk... Biological diversity – the variability of life on earth - is the key to the ability of the biosphere to continue providing us with these ecological goods and services and is thus our species assurance policy.
> (*How the Convention on Biological Diversity Promotes Nature and Human Well-being*, Secretariat of the Convention on Biological Diversity, p.1)

Biodiversity exists among different species and within each species. Differences within the gene pool for a single species are critical to the capability of a species to survive a change in its environment. Species exist in a variety of ecosystems such as deserts, forests, wetlands, mountains, lakes, rivers, and agricultural areas. There a number of species including humans who live in each of these different ecosystems. The Secretariat of the CBD reports that:

> It is this combination of life forms and their interactions with each other and with the rest of the environment that has made Earth a uniquely habitable place for humans. Biodiversity provides a large number of goods and services that sustains our lives.
> (*How the Convention on Biological Diversity Promotes Nature and Human Well-being*, Secretariat of the Convention on Biological Diversity p. 2)

Some of the goods and services which biodiversity provides include:

- Food and fodder;
- Nutrient cycling;
- Maintenance of hydrological cycle;
- Assures survival of regions through adaptation mechanisms;
- Provides food;
- Provides timber;
- Regulates climate, floods;
- Source of many medicines;
- Detoxification and decomposition of waste.

Human actions have drastically transformed ecosystems which have either been exploited for profits from forest products or have been burnt to produce a small area for farming. As with water, we destroy the forests and other ecosystems at a great risk to our own ability to survive. Consider the following transformations to the ecosystem which have impacted biodiversity:

- Between 1960 and 2000, reservoir storage capacity quadrupled and as a result the amount of water stored behind large dams is estimated to be three to six times the amount of water flowing through rivers at any one time;
- Approximately 35% of mangroves have been lost in the last two decades;

- Twenty percent of coral reefs have already been destroyed and another 20% degraded in the last several decades;
- Species have been disappearing at 50 to 100 times the normal rate;
- Based on current trends, 34,000 plant and 5,200 animal species will face extinction;
- Thirty percent of the main breeds of farm animals are currently at high risk of extinction;
- Forty-five percent of the world's original forests have been destroyed;
- Thirty-two percent of amphibians are threatened with extinction;
- There are 50 dead zones in the world's oceans.

(*How the Convention on Biological Diversity Promotes Nature and Human Well-being,* Secretariat of the Convention on Biological Diversity)
(*Human impact Triggers Massive Extinction,* Environmental News Service)

Biodiversity is so much more than the extinction of Bengal Tigers. It is actually an essential part of the services provided by the ecosystem to sustain life.

The United Nations Food and Agricultural Organization (FAO) released a major study in 2005 which assessed the sustainability of forests worldwide. According to the report, the United States has the seventh largest annual loss of primary forests and from 2000 – 2005, the U.S. lost an average of 215,200 ha or 531,771 acres of old-growth forests. Conversely, if plantations are taken into consideration, the United States gained 159,000 ha. The FAO concludes that primary forests are losing ground to modified natural, semi-natural, and plantation forest. (*Global Forest Resources Assessment 2005: Progress toward Sustainable Forest Management,* UN Food and Agricultural Organization)

Highest Average Annual Deforestation (2000 – 2005)

Country	Highest annual deforestation of primary forests 2000-2005
Brazil	-3,466,000*
Indonesia	-1,447,800
Russia	-532,200
Mexico	-395,000
Papua	-250,200
Peru	-224,600
U.S.	-215,200
Bolivia	-135,200
Sudan	-117,807
Nigeria	-82,000

*ha

(*Global Forest Resources: Progress Toward Sustainable Forest Management*, UN Food and Agricultural Organization)

Reforestation consists of planting new trees which constitutes 3.8% of total forest area or 140 million ha annually. Productive forest plantations, planted for wood and fiber production, account for 78% of forest plantations while 22% of forest plantation is planted to preserve soil and water. (UNEP) Clearly, replanting strategies need to be shifted so that more trees are grown to preserve soil and water.

Collectively, we have a responsibility to eschew our lifestyle choices to reduce the demand for resources and energy. Currently, burgeoning demand for products that we do not really need is depleting our resources including water, producing more waste than we can manage and spewing out toxic chemicals into the ecosystem. Some of us will pay the price now through health problems, but future generations are going to pay the heaviest price through externalized costs. We are passing on the costs for repairing the damage we caused to our children and grandchildren. We can only transfer the costs of growth for a limited period of time before it explodes in our face.

CHAPTER SEVEN

Muck Up:
Hazardous Waste and our Health

Love Canal was under construction in the 1880s to connect the upper and lower Niagara River until the construction company ran out of money and sold it at public auction. The Hooker Chemical Company, a subsidiary of Occidental Petroleum, bought the canal and from 1920 to 1953 dumped over 200 toxic chemicals at the site weighing 20,000 tons including DDT and other pesticides, solvents, PCBs, dioxin and heavy metals.

After filling the canal with dirt, the Hooker Chemical Company sold the property to the Niagara Falls Board of Education for one dollar. Incredibly, in 1954, the Board built an elementary school near the perimeter of the canal and by 1978, there were approximately 800 single-family homes, 240 low-income apartments and 400 children in the school near the site.

After reading a newspaper article in the local newspaper about the toxic dump-site, Lois Gibbs, a home-maker and mother with two children with recurring illnesses, began to question whether there was any connection between her children's poor health and the dump site. She met with the scientist who had been the source for the article and was handed a study outlining the risks associated with the chemicals in the dump.

Extremely alarmed after reading the study, she requested that her son, who was in kindergarten at the school, be transferred to another school, but the board refused. Subsequently, she began knocking on doors with a petition to shut down the school. While going from house to house, she discovered to her horror, that every home in the area had at least one member of the family who had suffered from birth defects, cancer, miscarriages, stillbirths or urinary tract infections.

Lois Gibbs and her neighbors formed the Love Canal Homeowner's Association. Their petition drive attracted media coverage and that, along with pressure from the Homeowner's Association and the residents, forced the New York State Department of Health to investigate the problem. They ordered the school to be closed, pregnant women and children under the age of two be evacuated and a cleanup undertaken.

All families were in a state of panic about the health risks and demanded that the Governor evacuate 239 families. However, there were six-hundred and sixty families still stuck in Love Canal and they pressured the State to pay for their relocation. Finally, the Federal government evacuated the remaining families. On October 1, 1980, President Carter visited Niagara Falls and signed a bill authorizing funding for any residents choosing to leave.

Subsequently, an informal study found that 56% of children born between 1974 and 1978 suffered birth defects, the miscarriage rate increased 300% and urinary-tract disease also increased by 300%.

Tragically, there are many love canals today although the main cause of health problems is now no longer dump sites. Rather, the enemy now is toxic chemicals spewed into our water, air and food. Manufacturing processes frequently generate toxic substances as a by-product which need a safe method for disposal. In the era of growth at any price, the cost of disposing of these hazardous by-products is not borne by the manufacturer. Instead, it is externalized so that either future generations pay the price for toxic waste or the current generation suffers the ill-effects.

Still in 2009, it's hard to believe that the costs of these harmful products are either invisible or ignored. Ironically, the very same people who are responsible for ignoring the costs are just as vulnerable to the ill-effects as everyone else.

In the United States alone, 4.1 billion pounds of toxic wastes were reported to the EPA for their Toxic Release Inventory Program (TRI) and 87% were released on site as follows:

- 32% (1.31 billion pounds) as air emissions;
- 19% (771 million pounds) in surface impoundments other than hazardous waste;
- 14% (590 million pounds) in underground injection wells;
- 14% (585 million pounds) as other land disposal (waste piles);
- 6% surface water discharges;
- .5% as land treatment;
- 14.5% misc.

(*U.S. EPA Toxic Release Inventory: Reporting Year 2007 Public Data Release*, EPA)

These hazardous substances are responsible for birth defects, miscarriages, respiratory illnesses, cancers and many other health-related issues. People world-wide are exposed to these toxins and shockingly, scientists have discovered dioxin, the most deadly chemical produced by humans, in the fat cells of polar bears. According to the United Nations Environment Program (UNEP):

> Air pollution caused the premature deaths of an estimated 70,000 people a year in the United States and 5,900 in Canada in the early 2000s and it is known to exacerbate asthma, which is on the rise, especially among children.
>
> (*North America: Global Environment Outlook*, UNEP)

The World Health Organization produced a report in 2009 which claims that:

> More than 25% of the global burden of disease is linked to environmental factors, including chemical exposures. For example, about 800,000 children each year are affected by lead exposure, leading to lower intelligence quotients... Worldwide, lead exposure also accounts for 2% of the ischaemic heart disease burden and 3% of the cerebrovascular disease burden... Some 9% of the global burden of lung cancer is attributable to occupation and 5% to outdoor pollution.
> (*Strategic Approach to International Chemicals Management*, World Health Organization)

Chemicals in the atmosphere are stored in the fat cells in our bodies. A Canadian study conducted by Environmental Defence reveals the extent to which we are contaminated by a mixture of toxic chemicals. According to the study:

> A cocktail of harmful toxic chemicals was detected in every person tested in a cross-Canada study of pollution in people conducted by Environmental Defence. *Toxic Nation: A Report on Pollution in Canadians* confirms that, no matter where people live, how old they are or what they do for a living, they are contaminated with measurable levels of chemicals that can cause cancer and respiratory problems, disrupt hormones, and affect reproduction and neurological development.
> (*Toxic Nation: A Report on Pollution in Canadians*, Environmental Defence)

The results of the study are alarming because contamination of the people who took part in the study is a harbinger for the level of toxins stored in the bodies of all people in North America. The study reports that:

Number of chemicals detected in the Toxic Nation Study

Chemical Group	Total Number of Chemicals Tested For	Total Number of Chemicals Detected	Average Number of In a Volunteer
Heavy metals	19	18	17
PBDEs	5	5	3
PCBs	16	14	10
PFOS	1	1	1
Orgnochlorine pesticides	13	10	8
Organophosphate insecticide	6	5	4
VOCs	28	7	1
TOTAL	88	60	44

(*Toxic Nation: A Report on Pollution in Canadians,* Environmental Defence, p. 15)

These chemicals only represent a fraction of the chemicals stored in our bodies because they were the only ones under study. We are all walking around with a myriad of dangerous chemicals stored in our fat cells with the certainty that some people will suffer ill-effects. We are all potential victims of the growth and profit practice of business and government. The above chemicals have the following ill-effects:

Number of chemicals detected in Study Volunteers that are Linked to Health Effect

Chemicals Effect on Health	Number of Chemicals Detected		
	Total	Average	Range
Carcinogen	41	28	18-36
Hormone disruptor	27	18	13-24
Respiratory Toxin	21	15	12-18

Reproduction/	53	38	28-46
Developmental toxin			

(*Toxic Nation: A Report on Pollution in Canadians*, Environmental Defence, p. 15)

There were 41 carcinogens for which the volunteers were tested and the average number found in each volunteer was 28. Bearing in mind that only a fraction of the carcinogens among the total number of hazardous waste products were used in the test, we are all victimized by carrying an alarmingly large number of carcinogens stored in our body's fat cells.

Most people today attempt to avoid carcinogens and other dangerous substances by giving up smoking, reading labels on food products, buying organic food or avoiding products that they know are dangerous. Unfortunately, these measures are inadequate to fully protect people from dangerous chemicals entering their bodies. We all still must breathe the air and drink the water.

Sadly, there are numerous reasons why people are so complacent about all the chemicals in the atmosphere. One reason is they do not know that the temple which houses their soul is ridden with these chemical devils. Furthermore, there is the problem of feeling overwhelmed because there are so many risks to our health and there seems to be no means to effectively address the problem. Therefore, they choose to ignore the hazards. The demands of work, family, finances and the need to escape life's pressures leave people with no energy to devote to social causes.

Strategies abound to minimize the risks of this problem such as avoiding foods with unsafe ingredients or buying organic food; speaking to the manager of your supermarket about the fact that many of the foods on the shelves are loaded with dangerous ingredients; writing or emailing people in the food industry; or simply becoming active by joining a group. Again, this is only a partial solution.

Frustratingly, the problem could be partly alleviated if the government were to impose stricter standards on the quality of our food and effectively enforce them. While other countries have been able to reduce the harmful substances in the environment, disappointedly, U.S. and Canada rank very poorly when compared to other countries. Consider the following two studies which rank environmental performance:

Environment Performance Rank of OECD Nations

Environmental Performance Rank of OECD Nations	Environmental Performance Ranking
Turkey	1
Switzerland	2
Denmark	3
Germany	6
Sweden	8
Netherlands	10
France	18
United Kingdom	18
Canada	28
Belgium	29
United States	30

(*The Maple Leaf in the OECD: Comparing Progress Toward Sustainability*, David Suzuki Foundation, p. 2)

The second study was conducted by the Conference Board of Canada and ranks countries according to environmental performance:

Environmental Performance Rankings

Country	Environmental Ranking
Sweden	1
Finland	2
Norway	3

Switzerland	4
U.K.	5
France	6
Italy	7
Austria	8
Germany	9
Ireland	10
Denmark	11
Belgium	12
Netherlands	13
Japan	14
Canada	15
Australia	16
U.S.	17

(*How Canada Performs: Ranking the Environment*, Conference Board of Canada, p. 3)

Clearly, in the United States, growth is a higher priority than the environment; otherwise, the government would create a set of high standards for regulating pollution and strictly enforce them. Strong environmental policies are threatening to corporate America in a number of ways: expenditures for complying with high environmental standards would encroach on profits which are virtually sacrosanct; and reducing wastes or emissions would incur new costs to either retool production or install pollution control equipment.

On the other hand, the government has a responsibility to take action given that their mandate is to serve the public interest and not to cater to the needs of corporate shareholders. By neglecting to protect its citizens, the U.S. and Canadian governments are actually sacrificing people on the altar of profits. Theoretically, it is possible to calculate the value of human life based on the government's decision to neglect the environment by adding up the total costs incurred for protecting citizens from toxic chemicals and dividing by the number of people who die from these chemicals. Relegating the responsibility for environmental protection to the private sector has been an unconscionable failure proving that the mindset about the role of government must be reexamined. Government has a very positive role to play and only those who benefit from a lack of government action are in some sense winners in accumulating greater profits. (It is true that the government heavily subsidizes many industries so that there is no lack of action when it comes to corporations)

The extent of the problem can be understood by examining the different health problems caused by the toxic chemicals that are ultimately deposited in our fat cells. The EPA produces a study referred to as the Toxic Releases Inventory (TRI) by collecting toxic waste data on a voluntary basis from various industries. It is important to bear in mind that the data may be incomplete given its voluntary collection methodology.

Carcinogens, or cancer-causing drugs, are a group of drugs that are commonly found in our bodies. In the United States, one in two men and one in three women will develop cancer over their lifetime. Astonishingly, in 1994, a full six percent or 33,900 people developed cancer because of carcinogenic chemicals in the environment. (*Toxic Releases in the United States*, EPA, p. 8)

According to the TRI report for 2007, an alarming 835 million pounds of carcinogens were produced as a result of manufacturing processes. Five of the carcinogens with the highest volume include:

Carcinogens Released in the Highest Volume to Air and Water, 2007 (pounds)

Chemical	Total Air and Water Emissions
Formaldehyde	21,498,318
Acetaldehyde	11,367,412
Benzene	6,274,731
Dichloromethane	5,897,623
Trichloroethylene	4,524,268

(*Releases: Chemical Report*, EPA)

Other carcinogens in the environment include carbon tetrachloride, chloroform, dioxins, polycyclic aromatic hydrocarbons (PAHS), arsenic, beryllium compounds and cadmium compounds. (*Cancer and the Environment*, National Cancer Institute, p. 12)

Industries releasing the most carcinogens air and water include:

Industries Releasing Carcinogens to Air and Water 2004 (pounds)

Industry	Total Air and Water Emissions
Chemical and Allied Products	17,320,073
Paper and Allied Products	12,409,815
Lumber and Wood Products	6,085,273
Petroleum refining and Related Industries	5,331,550
Fabricated Metal Products	4,097,416

(*Toxic Releases in the United States*, EPA, p. 8)

Another group of chemicals, known as developmental and reproductive toxins, can potentially cause fetal death, structural defects such as cleft palate, heart abnormalities and neurological, hormonal or immune system problems. According to the TRI, a shocking 96 million pounds of developmental toxins were released into the air and water in 2004. One of the worst culprits is Toluene which occurs naturally in crude oil and released during the process of refining oil. It can also be found in paints, fingernail polish, and adhesives. Five of the worst developmental toxins released into air and water include:

Developmental Toxins Released in the Highest Volume to Air and Water, 2007 (pounds)

Chemical	Total Air and Water emissions
Toluene	41,918,311
Carbon Disulphide	8,936,303
Benzene	6,274,731

N-methyl-2-pyrrolidone   4,471,426
1,3-Butadiene   1,788,269
(*Releases: Chemical Report*, EPA)

Industries responsible for dumping toxins in 2004 include:

Industries Releasing the Most Developmental Toxins to Air and Water,
2004 (pounds)

Industry	Total Air and Water Emissions
Chemical and Allied Products	29,723,637
Rubber and Miscellaneous Plastic Products	15,026,397
Printing, Publishing and Allied Industries	12,445,438
Paper and Allied Products	9,626,332
Petroleum Refining and Related Industries	6,983,661

(*Toxic Releases in the United States 2004* EPA, p. 12)

Five of the worst reproductive toxins released into air and water
include:

Reproductive Toxins Released in the Highest Volume to Air and Water,
2007 (pounds)

Chemical	Total Air and Water Emissions
Lead Compounds	475,542,383
Carbon Disulphide	8,936,303
Benzene	6,274,731
1,3-Butadiene	1,788,296
Ethylene oxide	310,773

(*Releases: Chemical Report*, EPA)

Industries releasing the most reproductive toxins into air and water in
2004 include:

Industries Releasing the most Reproductive Toxins to air and Water, 2004 (pounds)

Chemicals	Total Air and Water Emissions
Chemicals and Allied Products	21,496,968
Rubber and Miscellaneous Products	8,935,031
Petroleum Refining and Related Industries	2,160,364
Paper and Allied Products	1,214,972
Primary Metal Industries	1,190,040

(*Toxic Releases in the United States 2004*, EPA p. 12)

These are just examples of the different effects caused by chemicals released into our air and water. One of the other groups of chemicals that are harmful to our health, neurotoxins, impairs central nervous system functions causing fatigue, irritability and other behavioral changes. Some of the more serious effects include damage to the nerves carrying sensory and motor information and as well, degenerative diseases. The TRI 2004 Report found that 826 million pounds of neurotoxins were released into the air and water.

Respiratory toxins can cause irritation, bronchitis and cancer. According to the TRI 2004 report 1.5 billion pounds of respiratory toxins were released into the air.

One of the more pernicious effects of toxic substances involves a group of chemicals called endocrine disruptors. Currently, there is a controversy about whether these chemicals actually disrupt the endocrine system and about the dosage necessary before these chemicals become a health risk as endocrine disruptors. Clearly wildlife has been affected by endocrine disruptors as evidenced by the birth of species with abnormal characteristics.

The endocrine system within our bodies consists of a number of organs such as the pituitary gland, the pancreas, the adrenals, the testes and hormones which secrete carefully measured amounts of chemical messengers. These chemical messengers are responsible for maturation, growth and development.

Endocrine disruptors are chemicals that mimic or interfere with these messengers resulting in a wide range of disorders and developmental abnormalities. They can affect hormone levels, alter the function of these hormones' control, interfere with the body's ability to produce these hormones and interfere with the ways these hormones travel through the body.

For example, estrogen is a hormone, secreted by the ovaries, which regulates the menstrual cycle, fertility and a healthy pregnancy. Estrogen is essential for the normal development of the fetus.

Some of the endocrine disrupting chemicals include insecticides, fumigants, fungicides, industrial detergents, resins and plasticizers.

The above chemicals are just a sample of the toxic substances that can affect our health. In essence, we are all living in Love Canal but we have no place to go as did Lois Gibbs and her fellow residents. Business's responsible for these chemicals do not have to bear the cost of preventing or resolving the problem. Those costs are passed on to us, the healthcare system and future generations who are subsidizing these companies with their health or lives. Unless you want to attach a dollar amount to human lives, there is no alternative but to eliminate these toxins.

Governments have not enacted sufficiently strict regulations and a system of severe penalties for those responsible for producing these chemicals. Ideology and growth account for the lack of action to protect the public whom they were elected to serve. Disturbingly, to date, decisions on social and economic policy have not factored human health or the environment into the economic equation and industries have virtually been granted carte blanche to externalize all costs that might interfere with profits.

In addition to endangering human health, these chemical toxins have caused severe damage to the ecosystem on which we depend to sustain life. Coral reefs, the oceans and wildlife are threatened by these toxins.

Coral reefs are crystalline structures consisting of calcium carbonate which are produced by living organisms. Stony corals are the predominant organism in reefs which secretes an exoskeleton of calcium carbonate serving as the building blocks for corals. Reefs are most commonly found in tropical waters but as well, exist in deep cold water.

Due to their complex nature, thousands of species of fish and invertebrates live in association with coral reefs in a symbiotic relationship. Although they normally are located in nutrient-poor waters, the process of nutrient cycling between corals and other reef organisms allow them to flourish.

Corals are home to a variety of fish and other organisms which coexist in a food chain. They are the richest marine ecosystem in species, productivity, biomass, structural complexity and beauty. Reefs are home to more that 25% of all marine fish and to an estimated 10% to 15% of the total worldwide catch.

Additionally, they form wave-resistant structures by absorbing the energy of breaking waves which transport nutrients to the reefs and clean them of waste matter. Their wave-breaking capability serves to protect the shoreline from erosion.

According to the United Nations Environment Program, chemicals that remain intact in the environment for a very long time, or Persistent Organic Pollutants (POPs):

> ...are released into the environment through many activities including agriculture, forestry and urbanization. POPs may accumulate in Reef ecosystems through local releases or long-range transport. Many POPs have insecticidal properties and are relatively more toxic to invertebrates than

vertebrates, and may be endocrine disruptors in fish and mammals. (*Convention on Coral Reefs*, UNEP, p. 2)

Toxic chemicals which also include detergents, paints, oil spills, fertilizers and radioactive waste, are only one of many causes of the destruction of coral reefs. Other threats to coral reefs include over-fishing, destructive fishing methods, unsustainable tourism, coastal development, coral bleaching, rising sea levels and ocean acidification from increasing levels of CO2 in the atmosphere.

Coral reefs are an integral part of the earth's ecosystem yet we are systematically destroying them to sustain economic growth. The World Wildlife Fund reports that 20% of the world's coral reefs have been destroyed with no prospect of recovery. In the same report, the Fund predicts that 24% of the world's reefs are in danger of destruction in the short term while another 26% is in danger of destruction in the long-term. (*State of the Coral Reefs of the World*, World Wildlife Fund, p. 7)

Some of the worst cases of reefs either destroyed or in danger include:

Endangered or Destroyed Reefs

Region	Destroyed Reefs (%)	Reefs at Critical Stage (%)	Reefs at Threatened Stage (%)
The Gulfs	65	15	15
South Asia	45	10	25
SE Asia	38	28	29
US Caribbean	16	56	13
East Antilles	12	67	17

(*State of the Coral Reefs of the World*, World Wildlife Fund, p. 9)

As the home of coral reefs, the oceans are a singular example of humans fouling their own nest. Comprising 70% of the earth's surface, oceans are home to 70% of the world's biodiversity. They are a major factor in weather patterns, provide 70% of the atmospheric oxygen, absorb most of the planet's carbon dioxide and are a critical component in the hydrological cycle replenishing fresh water through cloud formation.

Despite its essential role in the maintenance of life, humans are threatening the health of the oceans in a number of different ways. Oceans have suffered the same fate as the atmosphere in the sense that for many years they seemed to be an infinite resource which was invulnerable to human activity. We are now tragically observing the results of that misplaced assumption.

We have treated the oceans as an infinite dump site and polluted them with chemicals such as DDT and PCBs, other organic chemicals, heavy metals, oil spills and sewage. On average, over 600,000 barrels of oil have accidentally have been spilled each year over the past decade.

Another threat to the complex and delicate ecosystem in the oceans is the problem of over-fishing which has depleted 70% of the world's commercial fishing grounds. (*Oceans and Coasts*, United Nations Environment Program, p. 2) Marine mammals are also threatened by large-scale industrialized fishing factories.

As discussed previously, the loss of coral reefs means the loss of habitat and a source of nutrients for many organisms. UNEP reports the significant loss of ocean mammal life in the following report:

Decline of Ocean Mammals - 1996

Species Decline	Past Population	Recent Population
Blue Whale	200,000	2,000
Right Whale	200,000	3,000
Bowhead Whale	120,000	6,000
Humpback Whale	125,000	10,000
Sei Whale	200,000	25,000

Fin Whale	470,000	110,000
Northern Sea Lion	154,000	66,000
Juan Fernandez Fur Seal	4,000,000	600

(*Take Action: A Guide for You and Your Community*, United Nations Environment Program, CH. 12, p. 6)

A shocking loss but even more so if you consider the fact that every species is part of a complex web of interactions with other organisms which depend on each other for maintaining the health of the oceans.

Unsustainable growth has performed its reverse Midas touch on the Oceans by exchanging profits for damage to the planet. All actions which have been destructive to the oceans resulted from human activity. Although these activities contributed to a nation's GDP, it was at a very steep price, not truly reflected in conventional economic measurements.

The good news is that the international community has recognized the severity of the risks and the urgency of the problem of toxic pollution. One group of chemicals, POPs, is particularly troubling due to their resistance to biodegradability. In 2001, in Stockholm, a group of nations passed the Stockholm Convention on Persistent Organic Chemicals. According to the Convention:

> The Stockholm Convention on Persistent Organic Pollutants is a global treaty to protect human health and the environment from chemicals that remain intact in the environment for long periods of time, become widely distributed geographically and accumulate in the fatty tissue of humans and wildlife. Exposure to Persistent Organic Chemicals can lead to serious side effects including certain cancers, birth defects, dysfunctional immune and reproductive systems, greater susceptibility to disease and even diminished intelligence.
>
> (*Stockholm Convention on Persistent Organic Pollutants*,
> United Nations Environmental Program)

POPs wreak the most havoc on humans and animals because even in small amounts they can cause cancer, weaken immune systems and interfere with reproduction. This treaty is an important step in eliminating chemicals harmful to life.

Unfortunately, the United States has not ratified the treaty despite signing the original document. Ratification means implementing legislation must be enacted so that it has legal force in the U.S. The Bush administration blocked the legislation at the 11th hour and Obama has yet to correct this disastrous error.

Again, unsustainable growth rides roughshod over human and environmental concerns. The mindset of not interfering with the activities of corporations that generate the economic activity which promote growth, must be replaced with a total commitment to sustainability. By refusing to ratify a critically important convention such as the one on POPs, Americans remain part of the problem rather than becoming part of the solution.

Another indication that the U.S. remains part of the problem is its refusal to completely embrace the precautionary principle (PP). The precautionary principle is a result of the evolution of methodologies for protecting humans and the environment from human action. It follows the *Polluter Pays Principle* and the *Prevention Principle* (PP). UNESCO published an extremely significant report in which it sought to reach a common understanding of the meaning of PP. In this report it formulated the following definition of PP:

> When human activities may lead to morally unacceptable harm that is scientifically plausible but uncertain, actions shall be taken to avoid or diminish that harm. Morally unacceptable harm refers to harm to humans or the environment that is:
> - threatening to human life or health, or
> - serious and effectively irreversible, or
> - inequitable to present or future generations, or

- imposed without adequate consideration of the human rights of those affected.

(The Precautionary Principle, The United Nations Educational, Scientific and Cultural Organization, p. 14)

It is somewhat equivalent to the principle of *reasonable doubt* applied by juries in judging the guilt or innocence of the accused. In very simple terms, it means that if a human activity and, in particular, a new technology or manufacturing process, may cause harm to humans or the environment but the risk is uncertain as to its effects, then it must be adapted or set aside until that risk is quantifiable and acceptable.

A prime example of human actions ignoring PP was the American refusal to apply it in the development of genetically modified (GM) foods. When GM foods were first produced and sold to the public, there was strong opposition to its introduction without adequate testing. A number of countries banned it including the countries in the European Union and Canada. Since then there has been more acceptance as exemplified by Europe introducing its first GM food in 2003. Currently WHO and the United Nations Food and Agricultural Organization have decided that GM foods are safe although the long-term effects may not be known for years. Controversy still surrounds GM foods but clearly in the early years, the U.S. introduced them in violation of the *precautionary principle*.

It is clear that for many years, the dangers of toxic chemicals remain in the environment and took a back seat to economic growth. Still today, economic growth is detached from any notion of sustainability. The problems of unsustainabilty and growth are still considered separate issues. The proof lies in the decision by the United States to pursue fossil fuels such as coal in the Appalachians and oil in the Caspian Sea Basin as well as Canada's decision to open its tar sands to anyone interested in earning a profit from expensive, dirty oil.

CHAPTER EIGHT

Hit Up:
Exploiting Developing Countries for Cheap Resources and Labor

*O*ne of the ugliest genocides since World War II occurred in Rwanda where 800,000 people died in 1994. Its cause was not just an internecine conflict between Tutsis and Hutus but a complex labyrinthian confluence of many causes which were social, economic and political in nature.

Hatred between the Hutus and Tutsis was sown during the colonial era when various colonial powers favored the minority Tutsis thereby building up resentment among the Hutus. After the colonial era, the Hutus took power and initiated a pogrom against the Tutsis many of whom fled to other African nations such as Uganda.

The expatriated Rwandans living in Uganda formed an army which invaded Rwanda in 1990. Attempts were undertaken to accommodate both sides in a coalition government but a group of extreme Hutus absolutely refused to share power with the Tutsis. When the possibility of a coalition government seemed real, the extreme Hutus embarked on a campaign to eradicate the entire Tutsi population and as well, moderate Hutus, who were willing to participate in a coalition government.

Economic factors were critical to the outbreak of violence. Actions of the World Bank, International Monetary Fund (IMF) and American trade policies drove Rwanda into the depths of poverty laying off idle angry young men who were available and ready for an opportunity to vent their frustration, anger and hatred.

The economic crisis was precipitated by the collapse of the international coffee market which had been stabilized by the 1987 International Coffee Agreement fixing quotas on coffee exports to control prices. A guaranteed price for international trade in coffee ensured that coffee farmers in developing countries were able to sell their coffee at a price which guaranteed them a living wage. Otherwise one or more countries could flood the coffee market with large quantities of coffee driving the price down so low that farmers in poor countries were no longer able to compete.

Coffee prices collapsed when large coffee traders in the United States pressured the government to withdraw from the International Coffee Agreement. Subsequently, coffee producers in the wealthy countries flooded the market with coffee, driving down the price. Subsidies and farm credits in the United States protected coffee farmers from any damage.

Such a steep decline in coffee prices resulted in a loss of 50% of export earnings in Rwanda leaving coffee farmers without money to buy food. A widespread famine ensued.

Rwanda was forced to approach one of the two major international lending institutions, the IMF, for financial assistance to rescue the economy and save the farmers from starvation. The IMF only loaned money to countries when they accepted certain conditions ostensibly to strengthen their economic fundamentals. In fact, the real agenda of the IMF, which acted in the interest of wealthy nations, was to impose neoliberal policies on the debtor nations to ensure that repayment of the debt was a top priority and to create favorable investment and trade opportunities for nations in the North.

In the case of Rwanda, one of the more devastating conditions imposed by the IMF was a requirement that the government devalue the Rwandan franc. Devaluation of the franc reduced even further the money farmers earned for growing coffee. At the same time, the cost of imports was rising steeply which made food and consumer necessities too expensive for many Rwandans.

Devaluation of the franc triggered a round of inflation, the collapse of real earnings and ultimately poverty. The consumer price index increased from 1% in 1989 to 19% in 1991 and the external debt expanded by 34% between 1989 and 1992.

As a result of the IMF prescriptions for a healthier economy, the education and health system collapsed and the incidence of severe malnutrition surged upward very rapidly. As a result of the IMF conditions, coffee cost less to purchase in the wealthy countries and the IMF was in a position to dictate any terms it believed were appropriate. In addition to devaluation of the franc, the IMF required the privatization of state enterprises, dismissal of public employees and a hike in the price of electricity. The purpose of these conditions was to free up money so Rwanda would be able to make its debt payments.

Since the Rwandan economy did not seem to be improving after the first devaluation, the IMF, in its infinite wisdom, imposed a second devaluation in 1992 which lowered coffee prices another 25%. The second devaluation would be equivalent to a doctor administering a particular drug to a patient and after observing the patients worsening condition, prescribed a heavier dose of the same drug. It all makes sense when you understand IMF decisions in the context of its agenda which was to serve the interests of the wealthy countries.

Rwanda is not an exceptional case in terms of the impact of World Bank and IMF policies. These two international banks are run by the wealthy countries who serve their own interests.

Dependence on the two international lending institutions usually begins when a developing nation is approached by an "economic hit man", a term coined by John Perkins who himself served that function as described in his book *Confessions of an Economic Hitman*. According to Perkins:

> Economic hit men (EHMs) are highly paid professionals who cheat countries around the globe out of trillions of dollars. They funnel money from the World Bank, the U.S. Agency for International Development (USAID), and other foreign "aid" organizations into coffers of huge corporations.
>
> (*Confessions of an Economic Hit Man*, John Perkins, p. xi)

More specifically, these economic consultants meet with leaders of developing countries to propose a development project such as the infrastructure for a mine or oil rig. The purpose of these projects is to lay the foundation for a major corporation to exploit the resources of the developing country. Frequently, the foreign company has no financial obligation to contribute to the construction of the roads, airports, electrical utilities which are needed to extract the resources.

The entire operation is a collaboration between the World Bank, the oil companies and the dictator of the host country. The oil companies frequently do not have to pay for the infrastructure and earn huge profits from the resource while the dictators receive a small percentage of the profits and pocket some or all of the money. The companies channel most of the profits back to the home office and share very little of it with the host nation. The host nation invariably is very poor and in desperate need of development money.

Usurping the resources of developing countries reduces the cost of resources and affords the predatory companies the opportunity to sell the final product at a lower price and to earn greater profits. One of the reasons that so many products in Western countries are cheap is a result of exploiting developing countries for cheap resources and for cheap labor in manufacturing or assembling the final product.

To maintain a growing economy entails not only convincing consumers to buy products they really don't need but to reduce the price of those products so that they are more affordable.

Tragically there are, as usual, many hidden costs of these products that are either borne by people in the present or by future generations. In the present, one of the invisible or externalized costs is the fact that people in developing countries frequently become displaced and lose their means of livelihood. The externalized costs, which have to be paid in the future, are a result of the damage to the environment and also the problem of growing numbers of displaced people.

As major players in the collaboration, both the World Bank and the IMF were conceived before the end of World War II at Bretton Woods, New Hampshire. Its ostensible purpose was to support post-war construction and the World Bank's first loan was to France in 1947 for US$250 million.

Eventually, the World Bank was subdivided into a number of associated institutions such as the International Bank of Reconstruction and Development (IBRD) and the International Development Association (IDA). After the rebuilding of Europe, its main focus has been on long-term development projects for developing countries.

According to the IMF, its purpose is "to provide policy advice and financing to members in economic difficulties and also works with developing nations to help them achieve macroeconomic stability and reduce poverty."

Both Banks were built by the wealthy countries of the world and were designed to ensure that control remained in the hands of wealthy countries. The structure and voting system within each of these institutions guarantees that decision-making will serve their interests.

One of the two main reasons that the Banks are under the control of wealthy countries in the European Union and United States and Japan is explained by the voting methodology of their 186 members. Voting shares are based on a country's share of the global economy or, in other words, it receives one vote for every dollar of its GDP. Clearly, the wealthy countries' will have far more votes than developing countries. Consider the voting shares of the following countries or group of countries for both the IMF and the World Bank:

Voting Share of Member Countries – World Bank and IMF (2009)

VOTING SHARE OF MEMBERS

Members	IMF	IBRD	IDA
Canada	2.89%	2.78%	2.72%
Japan	6.02%	7.86%	9.64%
Saudi Arabia	3.16%	2.78%	3.32%
United States	16.77%	16.38%	12.15%
European Union	32.08%	28.55%	32.94%

(*Voting Shares at the IMF and the World Bank*, Global Policy)

Note that Japan, United States and the European Union cast 54.87% of the votes at the IMF so that developing countries would be easily outvoted on policy decisions. In the case of the IBRD, these same three sets of countries constitute 52.79% of the total votes and for the IDA, they comprise 54.73% of the votes.

The World Bank claims that all powers are vested in the Board of Governors and the Board is the ultimate decision-making body in the Bank. Board members consist of one representative of each member country but it is unequivocally obvious that the wealthy countries control the outcome of all votes.

Members of the Board of Governors only meet annually and delegate all its powers to the Executive Directors with some exceptions. They meet twice a week to oversee the Bank's business including approval of loans, new policies, country assistance strategies and borrowing decisions.

The IMF has virtually the same organization and structure as the World Bank with one major exception: the President does not have to be an American.

There is a gigantic catch in this management structure which explains how the wealthy countries maintain control. There are 24 Executive Directors, five of whom are appointed by the Board of Governors and represent the five largest shareholders (United States, Japan, Germany, France and the United Kingdom) while the other 19 directors represent the remaining 181 member nations.

Shockingly, the Executive Directors are even more prejudiced in favor of the wealthy countries due to the fact that the same weighting system of voting shares applies as in the Board of Governors. In other words, the five countries that are represented on the Executive Board have considerably more votes than the other 19 members representing 181 countries. These five appointed members have 37.38% of the votes or on average 7.48 votes each while the 19 elected members have on average, 3.30 votes each or .35 per country. Incredibly, the Executive Directors chose the President of the World Bank who must be an American. The President is appointed for a five year term, chairs meetings of the Board of Directors and is "responsible for the overall management of the Bank".

It is patently and flagrantly obvious that the structure and organization of the Bank will produce decisions that serve the interests of wealthy countries. Nations act in their own self-interest so that when approving loans and choosing the conditions for the loans, they will not be designed necessarily to benefit the developing country receiving the loan. A few more examples will illustrate this point.

Jamaica is a quintessential example of a developing country borrowing money from international banks with imposed conditions, the result of which was the virtual destruction of the economy and in particular, the agricultural sector.

Between 1972 and 1976, Michael Manly, Prime Minister of Jamaica, embarked on a populist and nationalist program by creating rural health schemes, food subsidies, literacy campaigns, free secondary and higher education, a national minimum wage, equal pay for women, sugar cooperatives and rent and price controls. Revenues for these new projects were financed through an increased tax on the American and Canadian bauxite companies and through deficit spending.

The tax on Bauxite severely angered the United States who then embarked on a destabilization campaign in Jamaica. Aluminum and bauxite processing were moved to new locations, strict sanctions were imposed, foreign capital inflow dropped and the CIA began supporting opposition parties to unseat Manley in the next election.

Manley held firm and was reelected in a landslide victory in 1976. The economy suffered as a result of the American campaign as interest rates skyrocketed, imports were in short supply and lay-offs increased.

Due to the failing economy, Manley was forced to approach the IMF for a stabilization loan to remedy the government's balance of payment crisis. The IMF loaned Jamaica US$75 million but with massive strings attached. Manley was reluctant to meet all the conditions but relented in 1978 and received a second loan for US$240 million.

Devaluation of the Jamaican dollar was one of the conditions imposed by the IMF along with a new tax on consumer goods. Manley was forced to reduce government expenditures, increase charges for government services and reduce wages of public employees. Price controls were lifted, a ceiling on wages was imposed and companies were offered a guaranteed profit.

All these conditions were detrimental to the well-being of the Jamaican people who were paying the price for the IMF loans by losing their jobs, losing wages, paying higher prices for consumer goods and losing government programs and services. On the other hand, the private sector benefited as they were guaranteed an annual profit and were no longer subject to price controls. By 1979, having complied with the IMF conditions, Manley received an additional US$428 million in loans.

By the end of Manley's eight years in office, average income was down 25% and the cost of living had skyrocketed by 320%. Dissatisfied with their declining standard of living and loss of jobs due to the IMF conditions, the Jamaican people voted for Edward Seaga as Prime Minister and his party took control of the government in the 1980 election.

Seaga supported the IMF conditions and cut back the public sector even further and dismantled the Special Youth Employment program which had served 10,000 young people. He set a limit on public sector wage increases despite an inflation rate of approximately 30%.

Various banks including the IMF were now more than willing to loan Jamaica money and Jamaica's debt grew to US$2.2 billion. Ultimately the debt would reach US$4.5 billion in 2003 and Jamaica had already paid back 17 times the amount it had originally borrowed in interest payments alone. The outrageously high debt forced the government to pay 52 cents out of every dollar in debt repayment.

One of the major sectors in the Jamaican economy affected by IMF conditions was the agricultural sector.

Farmers in Jamaica remained competitive as long as the government imposed tariffs on imported food, subsidized its farmers and held down the interest rates on loans to the agricultural sector.

It was very difficult for farmers to compete with food exported from Europe and the United States because of the heavy subsidies awarded to their farmers, availability of cheap loans and the mechanization of large farms.

Jamaica was forced to abandon these protections as part of the conditions for IMF loans. As a result, Jamaican farmers could no longer compete with imported foods and suffered a collapse of their markets. According to Eurostep, an NGO cooperating with the European Union in promoting development:

> In late August 1999 Phyllis March, a dairy farmer in Jamaica, could no longer sell all the milk her cows produced. Her cooler was overflowing and she had to throw away nearly one thousand litres of fresh milk. She is not the only farmer having difficulties selling their produce. Local dairy farmers are losing the battle against cheap imports, and especially against subsidized milk powder coming from the European Union... Over the years, local dairy farmers have had to throw away hundreds of thousands of litres of milk from their overflowing coolers, because most local processors use the cheaper imported milk powder instead.
> (*Dumping in Jamaica: Dairy Farming Undermined by Subsidized EU Imports*, Eurostep)

Global Exchange, a human rights social justice organization based in the United States, reports that:

> We visited potato farmer Jerry Harrison who spoke of the difficulty of getting a farm loan even at 22% interest for spring planting. Jamaica was self-sufficient in potatoes in 1982, but has since been undercut by foreign imports.
> (*Globalization: Its Effects in Jamaica*, Global Exchange)

As a consequence of the economic reforms imposed by the IMF, Jamaica is one of the most indebted and impoverished countries in the world. It is saddled with an onerous debt burden which is the fourth largest per capita in the world. It faces increased foreign competition, exchange rate instability, a huge trade deficit and high unemployment.

To create markets for American and European products, Jamaica was forced to adapt its economy to meet the needs of the wealthy nations of the world. More markets represented more growth so growth again was the driving force in the plundering of the Jamaican economy.

Ecuador is another example of a country devastated by the actions of oil extraction and loans from the International Banks. In this example, the oil companies literally stole the oil and when they were through bolted out of the country leaving behind toxic waste destroying much of Ecuador's potable water. Texaco recklessly disposed of extremely hazardous waste in 600 open, unlined pits that have wreaked havoc on the health of the people of Ecuador.

Following Texaco's discovery of oil in Ecuador in 1967, Ecuador became the second largest oil exporter in Latin America accounting for half the national budget. Oil production also became a magnet for international loans all of which contributed to a very high growth rate but ultimately drove the economy into near-bankruptcy as the debt increased from US$241 million in 1970 to UIS$16.6 billion by 2006.

Indigenous people living in the rainforest had been unaffected by Western civilization. They lived in a very unique rainforest, rich in biodiversity until they were insidiously and treacherously lured off their land. One of the tribes of indigenous people, the Huanorani, lived in Yasuni National Park which had been designated by the United Nations Educational, Scientific and Cultural Organization (UNESCO) as an UNESCO Biosphere Reserve in recognition of its biodiversity and cultural heritage.

The Summer Institute of Linguistics (SIL) were an evangelical missionary group from the United States who worked in collusion with the oil companies and allegedly received a huge grant from Rockefeller charities, organizations with close ties to the oil industry. (*Confessions of an Economic Hitman*, John Perkins, p. 167)

According to the Huaorani, when seismologists reported that a certain region probably had oil beneath the surface, SIL members approached the tribe and encouraged them to move with promises of food, shelter, clothes, medical treatment, and education from the evangelical missionaries. (ibid)

Another tactic involved the SIL donating food laced with laxatives to the Huaorani and magically appearing at just the right moment with the medicine to cure the diarrhea epidemic. SIL members were regarded almost as deities and could easily lure the indigenous people off their land. (ibid)

Once in possession of land where they had located oil, Texaco built a network of roads and a pipeline that ran the length of the country laying waste to the surrounding rainforest. In addition, they built refineries and drilled for oil in hundreds of wells.

As a result of Texaco's pursuit of oil profits, local communities were devastated, forests were hacked down and the rivers poisoned by extremely toxic wastes. Indigenous groups such as the Huaorani, Cofan and Secoyas whose population once were in the tens of thousands were reduced to a few hundred each.

Following seven years of dictatorship, on April 29, 1979, Jaime Roldós, was democratically elected as president. Roldós was a populist and a nationalist who was determined to protect Ecuador from servitude to either the United Stated, multinational corporations or the international banking system.

His agenda consisted of social reforms and believed that revenue from oil, Ecuador's greatest potential resource, should serve the interests of the people of Ecuador. He introduced the Hydrocarbons Policy as a measure to finance his social reforms and to change Ecuador's relationship to the oil companies. The Hydrocarbons Act would guarantee that the majority of funds from oil revenues would be used to serve the people of Ecuador and that the oil companies would receive a reasonable share.

Not only were the oil companies extremely disgruntled about the loss of revenue but also about the precedent the Hydrocarbons Act would serve to the many other countries victimized by oil exploitation. Lobbyists and public relations people working for Texaco sprung into high gear vilifying Roldós as another Castro and twisting the arms of people in Washington and Quito to pressure Roldós to backtrack on his decision to steal his own money.

All of Texaco's efforts were to no avail as the President dug in his heels against threats and intimidation from Texaco and the American government. For his ardent nationalism, Roldós suffered the same fate as Allende in Chile in 1973 and Torrijos in Panama a few months later. He died in a fiery plane crash on May 24, 1981. There is no absolute proof but strong evidence points to the CIA as the perpetrator as in the other two cases. (*Confessions of an Economic Hitman*, John Perkins, p. 183)

In the 1980s the price of oil dropped and interest rates increased so that Ecuador was paying more to the banks and earning less from oil revenues. At the same time, the IMF continued to loan money to Ecuador. By 1999, repayment of the debt consumed 50% of the government's budget and as a result, funding for public services such as education, health and social security was slashed.

In April 2000, the IMF approved a US$300 million loan despite the fact that the government was already strapped with a US$15 billion debt. The loan opened the door to even more loans including US$425 million from the World Bank.

As a precondition for the loan, the government had to sign an agreement with the IMF which stipulated that private companies must be allowed to build and operate the pipeline in Ecuador. In 2003, the IMF loaned Ecuador yet another US$205 million.

The repercussions of Ecuador's economic woes, most of which were a result of its external debt, were the privatization of state institutions, extreme cutbacks to social programs, unemployment, extreme poverty and disastrous environmental degradation. In addition, there was also government corruption and mismanagement. Current economic statistics are as follows:

- Poverty rate: 38%
- Unemployment rate: 8.7%
- Debt: 29.2% of GDP
- Inflation rate: 8.6%
- Debt: US$16.8 billion

Although the economy has been improving since 2006, serious problems remain. External debt is still a serious problem and the poverty rate is extremely high. The worst legacy, by far, is the environmental disaster which seriously threatens the health of the people of Ecuador, destroying a valuable and important rainforest and polluting much of the potable water.

Texaco's operations destroyed 2.5 million acres of rainforest in the Ecuadorian Amazon. Shockingly, Texaco dumped 20 billion gallons of highly toxic wastewater into the water while creating 600 poorly constructed open waste pits contaminated with some of the most dangerous chemicals on the planet such as benzene, toluene, arsenic, lead, mercury and cadmium.

Cancer rates are 40% higher than normal for men and 60% higher than normal for women. Incidences of cancer are four times higher in the region where Texaco drilled for oil than in the rest of the country. The London School of Hygiene and Tropical Medicine discovered eight different cancers in San Carlos, a town surrounded by Texaco wells including: bile duct, stomach, larynx, liver, melanoma, leukemia, lymphoma and cervical cancer. (*Ecuador: Oil exploitation and environment rape*, Axis of Logic, March 5, 2005)

Ecuador is a crystal clear archetype of a developing country that large corporations and international banks exploited by stealing its resources and in the process, caused extreme harm to people and the environment. Sustainability was not an issue for all three perpetrators. Damage to the environment and the people of Ecuador, the burning of fossil fuels from the oil extracted and the damage to the economy were all examples of unsustainable practices for the sake of economic growth.

Even when the World Bank has the right intentions, unsustainable growth rears its ugly head. Chad and Cameroon are two of the poorest countries in Africa but they are endowed with considerable natural resources including oil in Chad. A consortium of ExxonMobil and Petronius from Malaysia were very interested in extracting the oil in Chad and a set of loans was arranged from the World Bank, International Finance Corporation (IFC -private sector lending arm of the World Bank), export credit agencies in the U.S. and private banks. The main lending body of the World Bank, the International Bank for Reconstruction and Development provided $93 million in loans and the IFC another $200 million. (*Contracting Out of Human Rights: The Chad-Cameroon Pipeline Project*, Amnesty International, p. 26)

The loans were designated for development of Chad's Doba oil fields in southern Chad and the related infrastructure and construction of a 1,070 km pipeline to transport the oil from Chad through Cameroon to the Atlantic coast at Kribi. Construction of the pipeline was completed and oil began to flow in July, 2003.

Two risks threatened the project: the World Bank's concern that none of the money would be allocated for development and the unstable nature of the government in Chad. In a shift of priorities, the World Bank decided to impose a condition that a substantial portion of the Government's share of the oil revenues must be devoted to assisting the poor in what it labeled a "model" for oil in development projects.

This shift was prompted by the grievously unsuccessful loans for oil in the past for which the Bank had received extensive criticism. Previous loans for oil projects had resulted in destruction of the local environment, displacement of people and massive human rights violations when governments suppressed local insurgencies.

Tragically, the project was not a "model" for development in oil but a repeat of past patterns in which local people and the environment were sacrificed on the alter of the oil companies. Amnesty International reports on the abuses in Chad:

> According to local NGOs, the operation of the oilfields and pipeline have already led to human rights abuses against many poor farmers in the Doba region of Chad, who were denied access to their land, which ExxonMobil refused either to compensate them for or to return to them. Several villages have reportedly been denied access to their sole safe water supply and the Kribi fishermen who work off Cameroons coast had their livelihoods seriously threatened by the pipeline.
> (*Amnesty International Investigation: ExxonMobil-led Oil Deal Threatens Law, Justice Through World Bank*, Amnesty International)

An insurgency grew out of resistance to the government's collusion with the oil companies' to deprive people of their land and to fail to alleviate poverty with oil revenues. President Devy's response to the insurgency and outbreak of violence was to mount a relentless counter-insurgency to suppress any opposition to the government and oil companies.

President Devy diverted some of the loan earmarked for development to the purchase of arms to support the counter-insurgency. Initially, Paul Wolfowitz, President of the World Bank, threatened to suspend loan disbursements to Chad for violating an agreement between Chad and the World Bank in which Chad had agreed to use a majority of the funds for development. Devy threatened to halt its exports and Western governments pressured the World Bank to reverse its course as part of an effort to prevent the anti-West insurgency from seizing power. Wolfowitz backed down and reversed the suspension.

In yet one more example, growth trumps human rights. Oil and a pro-Western government in Chad were a higher priority than the people who lost their land in the West's pursuit of oil.

In all these examples, corporations pursued profits without the least respect for human rights or the environment. International banks and wealthy countries collaborated in the unsustainable pursuit of profits to generate growth in the wealthy countries and to support their corporations who only see dollar signs when you mention sustainability.

CHAPTER NINE

Ad Up:
Paying for Political Engagement: the
marketplace of political change

21st century capitalism has met its nemesis. When the major threats to modern civilization become popularly accepted as inbuilt consequences of the capitalist system, they can no longer be relegated to outside the logic of economic growth. Questions about the viability of continued capitalist growth are tied to concerns around specific consequences like the following: The growing disparity of wealth between the developed and developing world, toxic pollutants in the biosphere, universal ecological hazards, and scarcity for key resources. These dangers have obtained a distinct position within the collective imagination and given rise to questions for the role of big business as exclusively profit seeking entities. New expectations for the way corporations conduct their business are now observed with Corporate Social Responsibility (CSR) mandates, as consumers begin to exercise political discretion in the marketplace. This chapter will consider some of the recent conditions that have provided for the current normative relationship between corporate objectives and social ethics, and particularly the way a new model of consumer activism has helped to reconcile tensions previously characteristic of early nineties corporatization. The pervasiveness of ethical labels in the current marketplace will be deemed both a positive indicator of consumer concerns for sustainability and a limited frame-work for political action. While there are undoubtedly benefits to the provision and consumption of ethical brands, a look at the relationship between the marketing of ethics and actual practice can reveal some of the current gaps for market led activism.

Modern Threats

The dangers threatening contemporary modern society can, in part, be understood by considering how these 'new' threats differ from the previous realm of threats under pre-modern and early capitalist societies. Ulrich Beck's "Risk Society" draws a useful comparison, representing older notions of threat as tied up with catastrophes like plague, famine and war, which, though undoubtedly extreme in their consequences, eventually became subject to the control of modern rationality and made to constitute 'calculable risks'. Some degree of measurability and therefore predictability could be applied to earlier dangers, mainly as these were bound by space – the geographical areas affected – and time - the effects were thought to correspond with a particular known event that would eventually wane. Contemporary threats are, quite conversely, understood to accumulate indiscriminately with the modernizing process itself. They are characterized by incalculable and irreversible consequences that could impact on all life (humans, animals, plants) around the globe, but, for which there is no clear accountability. Knowledge about the source of modern threats is further said to be bound up with a heightened reflexivity for our roles in modern society. Beck uses the term "reflexive modernization" to describe the way a preoccupation with the possibilities of industrialization has been superseded by a dissemination of knowledge for its inherent risks (1992). Reflexivity is at the core of what we might now consider a culture of skepticism, felt in the way individuals and institutions are continuously engaged in negotiating their identities vis-à-vis unconventional paradigms of thought and action.

The beginning of popular recognition for global threats to modern civilization derives primarily from an environmentalist perspective. Particularly, the realization that the current standard of living for industrialized economies is simply not sustainable with the physical resources available has had a major role in distinguishing anxieties for the current consumerist climate. According to Crane and Matten, the now pervasive notion of sustainability continues to be commonly associated with an environmentalist perspective. The aim of sustainability in this context refers to the following:

> All bio-systems are regarded as having finite resources and finite capacity, and hence sustainable human activity must operate at a level that does not threaten the health of those systems. Even at the most basic level, these concerns suggest a need to address a number of critical business problems, such as the impact of industrialization on biodiversity, the continued use of non-renewable resources such as oil, steel and coal, as well as the production of damaging environmental pollutants like greenhouse gases and CFC's from industrial plants and consumer products.
> (*Managing Corporate Citizenship and Sustainability in the Age of Globalization,* Crane and Matten, p. 25)

Two other key frameworks used to articulate sustainability are tied to perspectives on the economic and social aspects of the current global economy. In terms of the former, comprehensive notions of economic sustainability take into account both businesses' ability to sustain itself economically over time, but also to support the particular economy in which it is embedded. Social sustainability, which ties in significantly with the economic, refers mainly to the concept of social justice and especially the issue of big business practices contributing to deepening inequality across the globe (Crane and Madden, 2007). Appeals for social justice are especially varied, addressing both domestic and global disparities rooted in class, gender and race.

All three of the above-mentioned perspectives on sustainability play a part in informing our ideas about the threats produced by global capitalism. Trying to pin down the sites at which ideas about sustainability frequently translate into an ethics for practice, however, requires an understanding for the prevalent trend of consumer activism. More specifically, because our ideas about more sustainable modes of capitalist production and consumption are now mainly articulated within the context of the marketplace, the code of ethics informing producers and consumers interested in addressing the problems of consumption follows (somewhat paradoxically) from a market logic. The now common practice of companies adopting CSR mandates reflects attempts by companies to align themselves with discourses around sustainability. Even with PR resting as the unequivocal incentive behind CSR, corporations have become major actors in initiating the discourse around a more ethical marketplace.

In "Managing Corporate Social Responsibility in Action: Talking, Doing and Measuring", Karolina Windell discusses the selling of the CSR managerial model to corporations as a "modern business strategy". She describes the rhetoric of consultants in convincing corporations of CSR's benefits to involve a differentiation between reactive and proactive approaches to managing increasing expectations for more socially oriented business. While reactive strategies were presented as a "risk minimizing" strategy and could serve to manage criticisms for disreputable practices, proactive strategies positioned CSR as providing for stakeholders interests and eventually enhanced profit. Specific indicators for the benefits of the proactive approach, and thus CSR, are put forward as follows:

- Increase license to operate
- Enhance reputation
- Value added brand
- Recruit the best employees
- Enhance morale
- Increase efficiency and profits
- Good business sense

(*Managing Corporate Social Responsibility in Action: Talking, Doing and Measuring, 2007,* Windell, K., p. 43)

From the Margins to the Centre: The Voice of Social Change

A necessary condition for the current model of consumer activism has been the collapsing of the dichotomy between big business and social progressivism. While activism on the part of consumers can be traced back to the beginning of industrialization, this normally involved people drawing on their power as consumers against violations committed by big business, both domestically with things like health and safety abuses, but also with regard to exploitative activities abroad. More recently, in the years leading up to a socially responsible model for business, we witnessed a variety of activism against big business that did not appeal to consumer rights and interests. Many of the defining progressive movements of the nineties deemed consumer culture immediately relevant to the problems and injustices caused by big multinational corporations and accordingly sought to assert themselves as global citizens. Their common call for change to the very structure of a global capitalist economy based on exploitation and deceit placed these movements in the margins of consumer society.

Much of the publicized critique by nineties grassroots organizations addresses a concern for the social and ecological costs of the corporate pursuit of profits, together with the ideological conditions sustaining this. While varied in their methods, ranging from aesthetic transgressions, most notably culture jamming – "the practice of parodying advertisements and hijacking billboards in order to drastically alter their messages" (Klein, No Logo, 280), to protests targeting the exploitative activities of large corporations and the international bodies regulating these, this is a variety of activism that has assumed an oppositional stance to the key players and beneficiaries of the global economy. The face of anti-consumerism in the nineties is tied to a few key developments, including:

• **Adbusters**:	Founded in 1989, this is a "global network of artists, activists, writers, pranksters, students, educators and entrepreneurs who want to advance the new social activist movement of the information age." Concerned

with issues around corporate disinformation, injustices related to the global economy and pollution of the physical and mental commons, Adbusters is mainly renowned for its use of culture jamming techniques as a means to disrupting the meaning for widespread commercial messages. (Adbusters, 2008)

- **The Anti-Globalization Movement** A highly decentralized movement that became coined with two internationally organized protests in the late nineties against the activities of institutions like the G8, the World Bank and International Monetary Fund (IMF), with regard to their role in establishing the terms of trade in the global economy. Common to those involved is a concern for increasing privatization on a global scale, the associated social and ecological devastation, and the increasing presence of corporate messages in everyday culture. (Klein, New Left Review, 2001)

- **Redefining Progress** A think tank founded in 1994 with the intention to "develop and promote economically viable, socially equitable, and environment sustainable public policy." The assumption behind what Redefining Progress terms "smart economics" is that conventional models of economic growth do not account for ecological and social costs. Their development of the Genuine Progress Indicator (GPI) as an alternative to the Gross Domestic Product (GDP) takes into account natural and social capital and has been used by governmental and non-governmental organizations worldwide since its introduction in 1995.

(Redefining Progress, 2009)

These are three organizations that can help to delineate the parameters for the recent thrust of anti-globalization / anti-consumerist critique and action. Though the broad scope and variegated nature of this movement has been a major criticism, specifically in the way this creates confusion amongst proponents and opponents alike for a common goal, they have had an arguably strong impact on our collective understanding of the issues tied to global corporate expansion.

The focusing of progressive politics onto corporatization in the nineties is often credited for the shift towards a more socially conscious role for private business. At least in part, this can be seen as a pragmatic response to boycotts and protests against companies in violation of human rights and the environment taking their toll. But this also coincides with the sweeping rise of brand marketing and lifestyle consumption. In particular, as brand equity became a highly esteemed indicator of a company's overall assets in the last decade of the 20th century, companies began to bolster the energies and resources invested into a meaningful brand identity. Paradoxically, the previous political scrutiny directed at corporations could also be seen to impart on them a lucrative opportunity in the way of branding. Once already perceived as important political actors, it is not a far cry for companies to absorb politics within their brand identity, only now in a way that aligns them with progressive change. In many cases, this is exactly what has happened.

We now find that some of the most pervasive campaigns for sustainable development are the result of these being absorbed within the marketing of brands for particular companies. This is an arguably positive step in terms of creating a politics of action that is accessible within consumer society. It nevertheless accepts a model of action in which political advocacy is a secondary consideration. When social and ecological progress is put in the framework of market exchange, it is transferred onto the dynamic of the producer and consumer relationship. The particular business objectives for a company will decide what type of social responsibility platform suits its market. Similarly, the individual needs and desires of consumers will decide how to direct their political participation. Ethical brands invite us to see consumption as a way to exercise political discretion and consumers do make choices based on politics. But this choice is also motivated by other key factors like price, product quality, brand familiarity and convenience (Drieson et al., 2005), which often overcome the desire to exercise a social conscience.

In addition, knowledge about the cause emphasized by a company's brand and trust in their methods for acting on these are also important determinants for the choice to go ethical (Nichols and Opal, 2005). As a result of this variability, it should be no surprise that there is a discrepancy between the perceived importance for political cause and actual buying behaviour.

Fairtrade labels - a category of products traded under more economically favourable conditions for producers in developing nations - began to gain considerable momentum as a political initiative in the mid nineties. It could be said that Fairtrade "dominated the ethical consumption discussion" (Strong in Nichols and Opal, 2005, 183). Yet, despite its rising impact on consumer sensibilities, The Canadian Centre for Policy Alternatives reveals that:

> "Fairtrade ... still accounts for less than 1% of the North American market. And even though fair trade coffee benefits over 350,000 farmers organized into over 300 co-operatives in 22 countries, these farmers are still selling most of their crop outside of the fair trade system, because not enough companies are buying at fair trade prices. And yet, despite their relatively tiny share of total production, the fair trade movement is attracting considerable attention."
>
> (Fair Trade vs. Free Trade: Principles of fair trade based on economic justice, human rights, Oct. 2004)

There are a few major ethical labels giving shape to the market for socially conscious consumers, including: Ecolabels, Fairtrade certified products, Organic food labels, Forest Certification and Marine Certification. These all speak to the issue of sustainable development, mainly with an emphasis on one aspect - the economic, environmental or social. However, they are generally not engaged by consumers only on this basis. Certified ethical labels are integrated into the wider brand of companies, often as a way to extend its value. Rather than buying into the principals of Fairtrade or Organic food labels, alone, we are often buying into the principals of the brand of which this is a part. This can have the effect of actually diverting attention away from the cause being invoked, especially when this becomes wrapped up in a marketing language. Moreover, our accustomedness with the blending of ethical ideas and marketing tactics has arguably made us more vulnerable to 'washing' – a term that describes a company's efforts to communicate a socially responsible image of itself that is betrayed by their actual practices.

Greenpeace identified a particular case of Green washing as the 2007 General Motors "Gas Friendly to Gas Free" campaign, which outlined five ways that its Chevrolet line was going green: "Increasing fuel efficiency; producing vehicle than can run on E85 ethanol; and developing hybrids, plug-in hybrids and fuel cells". They further reveal that the impact on the public's perception of GM (according to one survey) was considerably positive, with 51% of respondents professing a better opinion of GM than a year ago, 35% of which citing "green" technology as the reason. Despite the massive advertising efforts directed at a green image, "(s)uper-efficient vehicles represent a small percentage of the 9.3 million cars GM produced in 2007. The company is only now introducing real hybrid vehicles, and its electric and fuel cell technologies are not yet production-ready".

The line between a company 'washing' and putting to use normal marketing strategies can be a fine one. Companies today are mainly in the business of selling the meaning behind their brand, which is often achieved by tapping into familiar cultural ideas and myths. Without actually defining a company's activities as socially responsible, such an image can be produced through mental associations. Grant remarks how "(m)any brands are trying to use green cultural codes - for instance, certain key words or visual images – to suggest their greenness without actually saying it" (The Green Marketing Manifesto, 2007).

Problems for washing are directly bound up with ambiguity for the way CSR strategies come to bear on Corporate Social Performance (CSP). There is a discernible vagueness for communication of CSR's real impact, which is the result of unclear definitions and methods of measurement for CSR and CSP, in addition to inaccessibility for key information about CSR practices for external actors (Den Hond et al.,). This makes it difficult to distinguish between companies engaged in 'washing' and those that adopt genuine CSR initiatives and equally difficult to determine the various degrees to which CSR influences a company's practices. The effectiveness of consumer activism today thus relies significantly on the critical knowledge of the consuming public in directing their participation vis-à-vis the ethical marketplace. Not only being informed about the particular practices of a company that qualify it as socially responsible, but the standards set up by third party institutions issuing certification of such practices (when this applies). More than this, however, the critical knowledge of consumers needs to exceed the parameters of consumer activism for effective social and ecological progress to occur. We cannot work against the hazards confronting modern civilization by making a few more ethical choices in the marketplace. In other words, the political agency of individual should precede all behaviour if this is to remain consistently devoted to an ethics of practice. While the rise of CSR and popularity of ethical brands have certainly helped to bring important issues around sustainability to the fore, the potential of today's consumer activism essentially depends on the way this is integrated within the day to day activities of individual political agents.

CHAPTER TEN

Act Up:
How to Have an Impact

Throughout history, people have successfully triumphed over injustice, inequality, exploitation and tyranny. Democracy, women's rights, freedom and sovereignty have all prevailed after long, difficult struggles by those who have been victimized.

For example, women's and civil rights in the 1960s, the end of colonialism in the latter half of the 20th century and gay and lesbian rights have succeeded in chiseling away at the chains that have bound them to an inferior status.

The next major challenge is to free ourselves from the vice-like grip of corporate rule which is the almighty power behind unsustainable growth. It is highly unlikely that powerful interests and their servants in government will readily renounce their relentless pursuit of profits irrespective of the costs to humanity or the environment.

However, a struggle by the people for a sustainable way of life will succeed if people commit themselves to taking responsibility for the society in which they live. If such a proposition seems unrealistic and naïve, consider the alternative. The unrealistic option is living in a world that ignores the dangers of unsustainabilty. Throughout this book, those dangers have been clearly delineated in order to emphatically demonstrate that the prognosis is undeniably negative for humanity if sustainability is ignored.

There are many obstacles to overcome before people will take up the cudgels against unsustainable growth. People need to be informed of the certain dangers that lay ahead on the unsustainable path they have chosen. Once informed, people need to be persuaded that they are not helpless and powerless but rather hold all the power. They need to be urged to drag themselves out of their morass of denial and avoidance and be inspired to join the struggle.

One of the greatest obstacles facing potential participants in the urgent cause of converting society to sustainability is that people have no inkling of the powers they possess. Once people recognize these powers, they need to know how to participate in the movement for a sustainable society.

In convincing people of their ability to effect change, it is necessary to overcome their doubts about their own individual power triggered by the questions: how can I make a difference? I am only one person and therefore can make no difference? The flaw in this line of reasoning is that the doubt becomes a self-fulfilling prophecy. If every individual asked themselves the same question and then decided not to participate, nothing will ever change. On the other hand, if they convince themselves that they do have power and can make a difference and participate, something can and will happen.

Another flaw in the reasoning is the question: even if I do something, nobody else is participating and therefore what is the point in my taking some action? The flaw in this line of reasoning is that you are using other people's lack of commitment to justify your own lack of commitment. You are only responsible for your own behavior and not the behavior of others. In addition, every movement starts with a small number of people and ultimately gains momentum if the cause is shown to be important to people's lives.

People who have a reluctance to join a movement whose objective is to radically transform how government works should read what Henry David Thoreau, American author, poet, critic, historian and philosopher had to say:

> Must the citizen ever for a moment, or in the least degree, resign his conscience to the legislator? Why has every man a conscience, then? I think that we should be men first, and subjects afterwards. It is not desirable to cultivate respect for the law, so much as for the right. The only obligation which I have a right to assume is to do at any time what I think right.

Thoreau's words should be the watchword of any civic-minded, participating citizen who dreams of creating a truly sustainable society. Governments can err in their judgments and, in a democracy, it is the citizens who must hold it to account and force it to change. Therefore, it is not wrong to oppose your government when you think it has erred, it is your duty as a citizen.

Another excuse people invoke to justify not participating in their democracy is that they have family and work responsibilities and need leisure time. In fact, this provides an excuse for a lack of free time for political activism. My answer to them is that they can accomplish enough in ten minutes a week to make a huge difference.

For example, if every American emailed their Member of the House or Senate once a week, they would be strongly pressured into paying attention to their constituents. It is true that Members of Congress have many compelling pressures to contend with. There are lobbyists, members of interest groups, leaders of their own party and pressure from the White House but they cannot ignore a deluge of emails demanding that they refuse to vote any more funding for the occupation of Afghanistan or Iraq or that they vote to support a single-payer healthcare system.

Very few if any elected official will risk reelection no matter what other pressures he/she feels. One of the reasons that pressure groups can have so much clout with elected officials is that they can withhold their support in the next election. If a sufficient number of voters in their district threatened to vote for one of their opponents, you will have caught their attention. It is true that voters do not agree on a number of issues but on many of the important issues there is a strong majority who do concur. Single-payer healthcare is an example.

It takes about two minutes to discover how to contact your Senator or Member of the House. To find your member of the House:

- Log onto Google (or other search engine);
- Type in House of Representatives;
- At the top left-hand corner of the main web page it will give you the option of finding your representative by Zip Code;
- Enter your Zip Code and press enter;
- Click on the name of your Representative;
- On the bottom of her web page you fill find the option "contact form";
- Click on "contact form" and you will have an already formatted email.

You do not have to go into details. Simply state your opinion and any action you might take (re: elections) if any. The time spent is well less than ten minutes and you have just made a difference!

Reaching your Senator is just as easy:

- Log onto Google (or other search engine);
- Type in U.S. Senate;
- On the upper right-hand corner of the main web page it will give you the option to "find your Senator";
- Click on the downward arrow to obtain a drop-down list of States;

- Click on your State;
- The name of the junior and senior Senator will appear;
- Click on one of the two names;
- You are now on the Senator's main web page. On the very bottom click on "constituent services";
- On the Constituent Services web page, look under "related links" on the right hand side and scroll down to "contact me".
- Click on "contact me" and a formatted email will appear;
- Send your message.

Another way to have an impact is to join a group that is fighting for a cause you support. There are numerous environmental groups, veteran groups, anti-poverty groups, anti-war groups and groups fighting for the rights of children and adults all around the world.

One of the benefits of joining a group is that you support it financially with your membership fees. As well, the group will either mail or email you literature on a regular basis so that you can remain abreast of their latest campaigns and remain informed of the latest issues.

A truly important advantage of group membership is access to petitions online that simplifies the process of collecting signatures. Frequently, you receive an email from a group which gives you the option to sign a petition. The petition is online and you simply add your name plus a message if you choose. These campaigns can be very effective and the use of the internet creates the opportunity to collect a large number of signatures very quickly.

When you join a group, there are a number of benefits. For instance, financial support and increased membership are critical politically because the larger and more visible the group, the more likely that Members of Congress will pay attention. As the membership of a group grows, there is the potential for petitions with more clout.

As well, there are international groups with national chapters, American groups, and purely local groups. As an example of a purely local group, Lois Gibbs and the group in Love Canal were successful in forcing the government to relocate them. The success of this group is, in fact, not a rarity and demonstrates that commitment, patience and perseverance pay off.

The World Wildlife Fund (WWF) is an example of an international environmental group with national chapters. According to their website:

> WWF's mission is the conservation of nature. Using the best available scientific knowledge and advancing that knowledge where we can, we work to preserve the diversity and abundance of life on earth and the health of ecological systems.

They are involved in climate change, protecting the oceans, preserving coral reefs and preserving biodiversity. WWF is only one of hundreds of groups from which to choose based on your own interests.

There are also political groups, some associated with a political party and some not. Progressive Democrats of America is one such group that operates as a grassroots movement operating inside the Democratic Party whose mission is:

> To build a party and government controlled by citizens, not corporate elites – with policies that serve the broad public interest, not just private interests.

They are involved in a wide range of issues and frequently email you petitions to sign. It is important to note that you do not have to be a democrat to join this group.

Another path to having an impact is to transform your personal lifestyle. Collectively, Americans have the power to stop corporate irresponsibility in their tracks because profits are completely dependent on sales and, as a consumer, you can choose not to purchase products that are manufactured by companies that pollute, exploit labor or engage in other unethical practices. There are a number of organizations examining the concept of ethical practices and examining the practices of particular corporations.

There is an organization called Ethical Corporation which prints a newsletter ten times a year and hosts business conferences and training workshops. If you go on a search engine and key in "corporate responsibility" you will discover a myriad of organizations dedicated to corporate responsibility.

Wal-Mart is an example of a corporation that engages in a number of unethical practices such as exploiting cheap labor in other countries, exploiting workers who work in American retail outlets, discrimination against women and anti-union activities. People who live near or below the poverty line might not have a choice in whether to shop at Wal-Mart but those living comfortably can choose Wal-Mart as a target for a personal boycott.

Many fast-food restaurants also exploit their workers and sometimes the environment as well. For example, McDonald's is virulently anti-union and uses intimidation and threats to prevent workers from joining a union drive. They have been known to even close an outlet that was in danger of becoming unionized in order to stop the workers from succeeding.

Some fast food hamburger chains raise cattle in the Amazonian rainforest to save money, destroying part of the rainforest in the process. If you believe in animal rights you might consider a boycott of meat from factory farms where cattle, pigs, chickens and other animals are terribly abused.

These are but a few examples of how you can vote with your dollar and it has worked in the past although usually with organized boycotts. For example Nike experienced a temporary drop in sales for exploiting workers in developing countries. Nestlé also suffered a drop in sales when a worldwide boycott was organized to stop the company from selling powdered milk to new mothers in developing countries.

In addition to boycotting a company's products, you can write or email a message to the company explaining why you are not buying their products. It is important to inform companies that people are aware of their unethical practices to give them an incentive to reform.

Ultimately, consumers possess all the power in deciding what is produced and what is not. You do not need to feel compelled to boycott all products manufactured by companies who lack corporate responsibility. You can start with one product and add more to your list or not. Everything you do is a contribution and there is no minimum amount of effort that is required.

As well, there are a multiplicity of lifestyle changes that would benefit the environment. The same principles apply. Whatever you choose to do is a contribution and it does not matter what other people are doing. You may even have an influence on your family, friends and neighbors.

Changing our lifestyle and consuming habits is not an option unless you are totally indifferent to the fate of the earth and the people who inhabit it. It is highly unlikely that corporations and governments will take the initiative so that responsibility for change devolves on the individual citizen. At some point you will be forced to change your lifestyle. By participating in the solution, you have more choices. Otherwise, floods, climate change, toxic chemicals, lack of water or rising ocean levels will compel you to change not by choice but by force.

Bibliography

Adamson, G. (2005, April 1). *Industrial Design: How Brook Stevens Shaped your World.* Cambridge: The MIT Press.

Amnesty International. (2005, September). Contracting out of Human Rights: *The Chad-Cameroon Pipeline Project.* Retrieved from http://www.amnesty.org/en/library/assetPOL34/012/2005/en/76f5b921-11dd-8a23-d58a49c0d652/pol340122005en.pdf.

Amnesty International. (2005, September 7). Amnesty International Investigation: *ExxonMobil-led Oil Deal Threatens Law, Justice Through World Bank.* Retrieved from http://www.amnestyusa.org/document.php?lang=e&id=ENGUSA20050907001.

Axis of Logic. (2005, March 5). *Ecuador: Oil exploration and environment rape.* Retrieved from http://www.axisoflogic.com/artman/publish/printer_16093.shtml.

Barber, A. (2008, May 20). *Levels of Economic Insecurity Job and Quality in the States in the First Half of the 2000s.* Retrieved from: http://www.cepr.net/index.php/press-releases/press-releases/new-report-details-levels-of-economic-insecurity-and-jon-quality-in-the-states/.

Barlow, M. (2007). *Blue Covenant: The Global Water Crisis and the Coming Battle for the Right to Water.* Toronto: Mclelland & Stewart Ltd.

Beck, U. (1992). *Risk Society.* London: Sage Publications.

Boushey, H. (2002, March 2). *The Needs of the Working Poor.* Retrieved from: http://www.epi.org/publications/entry/webfeatures_viewpoints_boushey_testimony_20020214/.

Boushey, H., Fremstad, S., Rachel, G., and Waller, M. (2007, March). *Understanding Low-Wage Work in the United States*. From the website of The Centre for Economic and Policy Research: http://www.mobilityagenda.org/lowwagework.pdf.

California Energy Commission. (2008). *Understanding Entrainment at Coastal Power Plants: Informing a Program to Study Impact and Their Reduction*. Retrieved from http://www.ca.gov/2007publications/CEC-500-2007-120/CEC-500-2007-120.pdf.

Car-Accidents. (2008). *Car accident Statistics*. Retrieved from http://www.car-accidents/pages/stats.html.

Carey, J. (2009, March 5). *Obama's Cap-and Trade Plan*. New York: Business Week. Retrieved from http://wwwbusinessweek.com/print/magazine/content/09_11/b4123.

CIA World Fact Book, (2009, April23). *Country Comparison – GDP-Per Capita*. Retrieved from https://www.cia.gov/library/publications/the-world-factbook/

CIA World Fact Book, (2009, April23). *Field Study – Unemployment Rate*. Retrieved from https://www.cia.gov/library/publications/the-world-factbook/.

CIA World Fact Book, (2009, April23). *Infant Mortality Rates*. Retrieved from https://www.cia.gov/library/publications/the-world-factbook/.

Conference Board of Canada. (2008, October). *How Canada Performs: Ranking the Environment*. Retrieved from http://www.conferenceboard.ca/hcp/details/environment.aspx.

Council of Economic Advisors. (1953). *Annual Economic Review*. Retrieved from http://fraser.stlouisfed.org/publication/ERP/issue/1724/download/14227/ERP_AER_1953.pdf.

Council of Economic Advisors. (1961). *Annual Economic Review.* Retrieved from http://fraser.stlouisfed.org/publication/ERP/issue/1447/download/5617/ERP_ACEA_1961.pdf.

Council of Economic Advisors. (1965). *Annual Economic Review.* Retrieved from http://fraser.stlouisfed.org/publication/ERP/issue/1200/download/5651/ERP_ARCE_1965.pdf.

Council of Economic Advisors. (1973). *Annual Economic Review.* Retrieved from http://fraser.stlouisfed.org/publication/ERP/issue/1222/download/5695/ERP_ARCE_1973.pdf.

Council of Economic Advisors. (1978). *Annual Economic Review.* Retrieved from http://fraser.stlouisfed.org/publication/ERP/issue/1381/download/5768/ERP_ARCE_1978.pdf.

Council of Economic Advisors. (1986). *Annual Economic Review.* Retrieved from http://fraser.stlouisfed.org/publication/ERP/issue/1591/download/6013/ERP_ARCE_1986.pdf.

Council of Economic Advisors. (1995). *Annual Economic Review.* Retrieved from http://fraser.stlouisfed.org/publication/ERP/issue/1616/download/7483/ERP_ARCE_1995.pdf.

Council of Economic Advisors. (2007). *Annual Economic Review.* Retrieved from http://fraser.stlouisfed.org/publication/ERP/issue/2238/download/29627/ERP_ARCE_2007.pdf.

Crane, A. and Matten, D. (2007). *Managing Corporate Citizenship and Sustainability in the Age of Globalization.* Oxford: Oxford University Press.

David Suzuki Foundation. (2005). *The Maple Leaf in the OECD: Comparing Progress Toward Sustainability.* Retrieved from http://www.davidsuzuki.org/Files/WOL/OECD-EngExec.pdf.

Den Hond, F. et al. (2007). *Managing Corporate Social Responsibility in Action: Talking, Doing, Measuring.* Surrey (UK): Ashgate.

Department of Energy. (2008, July). *Renewable Energy Consumption and Electricity Preliminary Statistics.* Retrieved from http://www.doe.gov/creaf/alternate/page/renew_energy_consump/pretends.pdf.

Dickie, P. (2007, June). *Making Water Desalination: Option or distraction for a thirsty word?* World Wildlife Fund. Retrieved from http://wwf.org.au/publications/desalinationreportjune2007.

Drieson, L. et al. (2005). *Do consumers care about ethics? Willingness to pay for fair-trade coffee,* Journal of Consumer Affairs, winter. (accessed 2009-08-28)

Environmental Defence. (2005, November0. *Toxic Nation: A Report on Pollution in Canadians.* Retrieved from http://www.toxicnation.ca/files/toxicnation/report/PATIN_English.pdf.

Environmental News Service. (1999, August 2). *Human Impact Triggers Massive Extinctions.* Retrieved from http://www.ens-newswire.com/ens/aug1999/1999-08-02-06.asp.

Environmental Protection Agency. (2004, July 9). *National Pollutant Elimination System.* Retrieved from http://www.epa.gov/EPA-GENERAL/2004/July/Day-09/g4130a.htm.

Environmental Protection Agency. (2007). *U.S. EPA Toxics Release Inventory: Reporting Year 2007.* Retrieved from http://www.epa.gov/tri/tridata/tri07/pdr/key_findings_viza.pdf.

Environmental Protection Agency. (2009, May 14). *Releases: Chemical Report.* Retrieved from http://www.epa.gov/cgi-bin/broker?view=USCH&trilib=TRIQ1&sort_fmt=18tstate8col.

Eurostep. (1999, November). *Dumping in Jamaica: Dairy Farming Undermined by Subsidized EU Imports.* Retrieved from http://wwweurostep.antenna.nl/detail_pub.phtml?page=pubs_position_coherence_jamaicad.

Friedman, M. (1962). *Capitalism and Freedom.* Chicago: University of Chicago Press.

General Accounting Office. (2003, July). *States' Views of How Federal Agencies Could Help Them Meet the Challenges of Expected Shortages.* Retrieved from http://www.gao.gov/new.items/d03514.pdf.

Global Policy. (2009). *Voting Share at the IMF and the World Bank.* Retrieved from http://www.globalpolicy.org/component/content/aericle/104/46584.html.

Grant, J. (2007). *The Green Marketing Manifesto.* Etobicoke (ON): John Wiley and Sons

GreenPeace. (NA). *Gas-Friendly to Gas-Free?: GM's Attempt to Greenwash its Image.* Investigation Brief. (accessed 2009-09-10).

Institute for Research on Poverty. (2009, March 19). *What are the poverty thresholds* and poverty guidelines? Retrieved from http://www.irp.edu/faqs/faq1.htm.

International Energy Agency. (2009). *Renewables Global Status Report/2009.* Retrieved from http://www.iea.org/files/Renewables_Global_Status_report.pdf.

Kennan, G. (1948). "Review of Current Trends in U.S. Foreign Policy". *Foreign Relations of the United States.* Volume 1; p. 509-529.

Klein, N. (2000). *No Logo.* Toronto (ON): Random House of Canada.

Klein, N. (2001). *Reclaiming the Commons.* New Left Review 9, May-June.

Labor Research Association. (N/A). *Wages and Benefits: Real Wages (1964-2004)*Retrieved from http://www.workinglife/org/wiki/Wages+and+Benefits%3A+Real+Wages+(1964-2004).

Makhijani, A. (2009, May). *Nuclear Power and CO2 Emission Reductions.* Retrieved from the Institute for Energy and Environmental Research Website http://www.eer.org/carbonfree/Nuclear Power Wastes and co2 copst reduction considerations.pdf.

Meadows, D., M., Meadows, D., L., Randers, J., and Behrens III, W., W. *The Limits to Growth.* New York: Universe Books

National Academy of Sciences. (2007, February). *Colorado River Basin Water Management: Evaluating and Adjusting to Hydroclimatic Variability.* Retrieved from http://www.dels.nas.edu/rpt briefs/colorado river management final.pdf.

National Centre for Children in Poverty. (2009, June). *Ten Important Questions about Child Poverty and Family Economic Hardships.* Retrieved from http://www.nccp.org/publications/pub 829.html.

Nichols, A. and Opal, C. (2005). *Free Trade: Market Driven Ethical Consumption.* London: Sage Publications.

OECD. (2006). *Science Competencies for Tomorrows World: Volume 1 – Analysis.* Paris: OECD Publishing.

Porter, E. (2006, October 15). *After Years of Growth, What About Workers Share?* New York Times: Retrieved May 15, 2009 from http://www.nytimes.com/2006/10/15/business/yourmoney/15view.htm.

Reich, B., Robert. (1999, March 29). Time100. *John Maynard Keynes.* Retrieved from http://www.time.com/time/time100/scientist/profile/keynes.html.

Secretariat of the Convention on Biological Diversity. (2000, April). *How the Convention on Biological Diversity Promotes Nature and Human Well-being.* Retrieved from http://www.cbd.int/doc/publications/cbd-sustain-en.pdf.

State of the Union Address Library. (2009, February 24). *State of the Union Address.* Retrieved from: http://stateoftheunionaddress.org/.

Statistics Canada, (2007, December). *Low Income Cutoffs for 2007 and Low Income Measures for 2006.* Retrieved from http://www.statcan.gc.ca/750002m2008004-eng.pdf .

Talberth, J. (2009, February). *Economic Feasibility of Coal to liquids Development in Alaska's Interior.* Retrieved from the Centre for Sustainable Economy Website: http://www.sustainable-economy.org/main/news/1.

Talberth, J., Cobb, C. and Slattery, N. (2006). *The Genuine Progress Indicator.* Redefining Progress: Retrieved on May 12, 2009 from http://www.rprogress.org/publications/2007/GPI2006.pdf.

Tarnoff, R. (2004) *Fair Trade vs. Free Trade: Principles of fair trade based on economic justice, human rights.* Canadian Centre for Policy Alternatives, October, (accessed 2009-08-28).

The Wall Street Journal. (2009, March 9). *Who Pays for Cap and Trade?* New York: The Wall Street Journal. Retrieved from: http://online.wsj.com/article/SB12365559060906602.html#printMode.

The World Health Organization. (2000). *The World Health Organization Ranking of the Worlds Health Systems.* Retrieved from http://www.photius.com/rankings/healthranks.html.

United Nations Environment Program. (1996). *Take Action: A Guide for You and Your Community*. Retrieved from http://www.nyo.unep.org/action/12.htm.

United Nations Environment Program. (2003, May). *Conventions and Coral Reefs*. Retrieved from http://www.unep.org/PDF/Conventions_Coral_Reefs_optimized.pdf.

United Nations Environment Program. (2007, October 25). *North America: Global* Environment Outlook. Retrieved from http://www.unep.org/geo/geo4/media/fact_sheets/Fact_Sheet_12_North_America.pdf.

UNEP. (2008). Kick the Habit: A UN Guide to Climate Neutrality. Retrieved from http://www.UNEP.org/publications/Kick-The-Habit/pdfs/KickTheHabit_en_lr.pdf.

United Nations Educational Scientific and Cultural Organization. (2005, March). *The Precautionary Principle*. Retrieved from http://unescbc.un.unesco.org/images/0013/001395/139578.pdf.

UNICEF. (2005). *Child Poverty in Rich Countries: Report Card No. 6*. Retrieved from http://www.unicef.org/media/files/ChildPovertyReport/.pdf.

UNICEF. (2007). *Child Poverty in Perspective: An overview of child well-being in rich countries*. Retrieved from: http://www.unicef.org/Poverty_and_children_a_Perspective.pdf.

UNICEF. (2008). *State of the World 2008*. Retrieved from http://www.childinfo.org/files/The_State_of_the_World_Children_2009.pdf.

University Corporation for Atmospheric Research. (2009, April 21). Retrieved from http://www.ucar.edu/news/releases/2009/flow.jsp.

U.S. Bureau of Labor Statistics. (2005, February). *Contingent and Alternative Employment Arrangements.* Retrieved from: http://www. bls.gov/news-release/conemp.nr0.htm.

U.S. Bureau of Labor Statistics. (2007). *Occupational Earnings Tables – 2007.* Retrieved from: http://www.bls.gov/ncs/ocs/sp/nctb0298.pdf.

U.S. Bureau of Labor Statistics. (2008, August). *A Profile of the Working Poor, 2006.* Retrieved from http://www.bls.gov/cps/cpswp2006.pdf.

U.S. Bureau of Labor Statistics. (2009, February). *Employment Status of the Civilian Noninstitutional Population, 1940 to Date.* Retrieved from http://www.bls.gov/cps/cpsaat1.pdf.

U.S. Bureau of Labor Statistics. (2009, February). *How the Government Measures Unemployment.* Retrieved from http://www.bls.gov/cps/ cps_htgm.pdf.

U.S. Bureau of Labor Statistics. (2009, April). *Ranks of Discouraged Workers.* Retrieved from: http://www.bls.gov/opub/Ns/pdf/opbills74. pdf.

U.S Bureau of Labor Statistics. (2009, May). *Full-Time Civilian Workers.* Retrieved from: http://www.bls.gov/ncs/ocs/sp/nctb0313.txt.

U.S. Bureau of Labor Statistics. (2009, May). *Household Data: Annual Averages.* Retrieved from: ftp://ftp.bls.gov/pub/special.requests/1f/aat20. txt.

U.S. Bureau of Labor Statistics. (2009, May). *Full-time Civilian Workers.* Retrieved from: https://www.bls.gov/ncs/ocs/sp/nctb0314.txt.

U.S. Census Bureau. (2006). *Poverty Thresholds 2006.* Retrieved from http://www.census.gov/hhes/www/poverty/threshld/thresh06.htm.

Water Resources Research. (2006). *Updated Streamflow Reconstructions for the Upper Colorado River Basin*. Retrieved from http://www.colorado.edu/treeflow/docs/woodhouse_lees_ferry_wwr_2006.pdf.

World Health Organization. (2009, April 23). *Strategic Approach to International Chemicals Management*. Retrieved from http://apps.who.int/gb/ebwha/pdf_files/A62_19-en.pdf.

World Meteorological Society. (2005) *Climate and Land Degradation*. Retrieved from http://www.wamis.org/agm/pubs/brochures/wmo489e.pdf.

World Meteorological Organization. (2009). *Bulletin: World Climate Conference-3)*. Retrieved from http://www.wmo.int/wcc3/documents/58_3_en.pdf.

World Meteorological Organization. (2009). *Climate Information for Securing Food*. Retrieved from http://www.wmo.int/wcc3/documents/WCC3_factsheet4_foof_EN.pdf.

World Wildlife Fund. (2004, December 1). *State of the Coral Reefs of the World*. Retrieved from http://assets.panda.org/downloads/executivesummary01dec04.pdf.

World Wildlife Fund. (2007, June). *Making Water Desalination: option or distraction for a thirsty world?* Retrieved from http://wwf.org.au/publications/desalinationreportjune2007.

www.ingramcontent.com/pod-product-compliance
Lightning Source LLC
Chambersburg PA
CBHW020430290526
45785CB00002B/787